Geometry in Nature

Geometry in Nature

Vagn Lundsgaard Hansen

Mathematical Institute
The Technical University of Denmark
Lyngby, Copenhagen

Translated by Tom Artin

A K Peters
Wellesley, Massachusetts

Editorial, Sales, and Customer Service Offices
A K Peters, Ltd.
289 Linden Street
Wellesley, MA 02181

Library of Congress Cataloging-in-Publication Data

Hansen, Vagn Lundsgaard.
[Geometriske dimension, English]
Geometry in nature / Vagn Lundsgaard Hansen.
p. cm.—
Includes bibliographical references and index.
ISBN 1–56881–005–9
1. Geometry. 2. Nature. I. Title.
QA445.H3513 1993
516-dc20

92-16734
CIP

Printed in the United States of America

97 96 95 94 93 10 9 8 7 6 5 4 3 2 1

Contents

Preface to the English Edition

I AM PLEASED THAT my book on the role of geometry in our perception of the world is now available in an English translation. *Geometry in Nature* should provide not only scientists but also the educated layman with an insight into contemporary mathematics, and I hope thereby to demonstrate to a broader audience that mathematics is a subject very much alive. In fact, I planned the book so that our Danish Minister of Education would be able to read it (if he so wished).

Even the best translation may not accurately reflect all the special points originating in another language and culture. However, I do think that most of the original flavor of the Danish text has been preserved in the English translation, for which I am grateful.

Among the many colleagues and friends at the Technical University of Denmark who have been a constant source of inspiration and of great help to me in working on this book, I would like to single out Poul Hjorth, with whom I have had many useful conversations relating to geometry and physics. It is also my pleasure to thank Professor Sigurdur Helgason, who brought the news about my book to the US.

Finally, it gives me great pleasure to thank Dr. Klaus Peters for a very encouraging and efficient collaboration in connection with this publication.

Vagn Lundsgaard Hansen
Lyngby, May 1992

Preface

THE ITALIAN GALILEO GALILEI (1564–1642), founder of modern science, has already stated it clearly: The language of mathematics and the figures of geometry are the language and the symbols of the universe. Thus, in the essay, *Sidereus Nuncius* (*The Starry Messenger*), from 1610, he writes (freely translated):

> Natural philosophy is written in that great book that lies eternally spread before our eyes—I mean the universe. But we cannot understand it before we learn the language and learn to understand the symbols in which it is written. The book is written in the mathematical language, and the symbols are triangles, circles, and other geometrical figures, without whose help it is impossible to comprehend a single word, and without which we wander through a dark labyrinth in vain.

Wherever we turn our glance, we see lovely geometrical forms, on both the small and the large scale. Just think of the enchanting symmetries and graceful patterns we perceive in animals and plants, in fascinating sculptures, and in imposing buildings. We seldom think of these from a mathematical point of view as we move about. One of the objects of this book is to help the reader discover the geometric dimension in the world around us.

Number theory and geometry, with a millenia-long history, together constitute the original mathematical disciplines. It is fascinating that the regularities and connections once discovered by mathematics retain their validity; more fascinating still is the fact that new relations are continually being discovered. Since mathematics is also a logical structure in which one step follows from another, it is by now quite a challenge to come to be at the cutting edge; the mathematical disciplines have evolved to the point that it is a lengthy trek to reach their frontiers. This is why we rarely get to know 20th-century mathematics through the school math curricula, and as a result one has the impression that mathematics is complete and finished. This book seeks to show that throughout human history new dimensions have continuously been established in geometry, and that its evolution is far from complete.

Although individual chapters in the book can in principle be read independently, a sequential development of the subject can be detected here. Along the way, the demands on the reader's abstract mathematical skills will vary, so that though one may falter at some points in the presentation, fresh opportunities lie just around the corner. Of necessity, a few sections do require knowledge of college-level mathematics. It is probably inescapable that a book about geometry would require from the reader a certain appreciation of spatial relations and geometrical intuition; the payoff is that he will ascend to an appreciation of the higher geometric dimensions.

At the back of the book, a bibliography lists the most important sources for the material of each chapter. Included are also several other works in which certain subjects are further elaborated.

In the preparation of this book, I have received valuable critiques from a number of my colleagues at The Technical

University of Denmark. To be singled out are Jens Gravesen, Lars Gæde, Per W. Karlsson, Steen Markvorsen, Frank Nielsen, and Flemming Damhus Pedersen. I have also received helpful comments from my students Jan Kristensen, Peter Gross, Michael Mikkelsen, and Kim Sparre. At various points in its evolution, I have had many good discussions about the book with Christian Thybo. The support of the series editor, Professor Stig Andur Pedersen, has been of particular value to me.

Finally, I wish to thank Nyt Nordisk Forlag Arnold Busck A/S for its excellent collaborative effort in the publication of this book.

Vagn Lundsgaard Hansen
Kokkedal, June, 1989

1
GEOMETRIC FORMS IN NATURE

THE OTHER DAY I VISITED THE ROUND TOWER in Copenhagen. On my way to the top I followed the outer wall, since the ascent there is easiest and I am not as fit as I once was. Czar Peter the Great did it with ease; he rode to the top! Well, my reward was the leisure for a little mathematical reflection on the Round Tower.

The surface one follows on the trip to the top of the Round Tower is a so-called *helicoid.* If we wished, we could pave it with straight boards radiating from the mid-axis of the tower, for it is a so-called *ruled surface*, i.e., it can be swept out by a straight line moving in space; and yet, the surface is obviously curved! Furthermore, the surface has the property that locally it displays minimal area. The helicoid is a unique surface in physical space that can be constructed with straight lines and locally displays minimal area. The latter property, incidentally, lends the surface particular strength, so that it has of course attracted the attention of civil engineers.

By the way, it was also fun to follow the wall for thereby I experienced firsthand taking a course along a helix. On a smaller scale, a helix is precisely the curve followed by a twining plant when it twists itself up a round stick. The helix is an interesting mathematical curve; it is the shortest path on the

Figure 1.1. Nautilus. *Photo A. Brovad.*

surface of a cylinder between two closely neighboring points, for example. It is thus the path an intelligent fly would take between two sweet splatters on the surface of a tin can.

In this chapter, we shall encounter all these things and more; and incidentally, my neighbor will find out why he cannot tile his terrace with pentagons.

1. Spirals and the Wonderful Snail

Who has not admired a snail shell or some other shell so beautifully constructed by nature? An especially elegant spiral is found in the shell of a primitive cuttlefish called the *nautilus*, shown in Fig. 1.1. What kind of curve is this? While we consider the nautilus as a model for our discussion in this opening section, the truth is that wherever we cast our glance, we find beautiful geometrical forms in nature, an experience

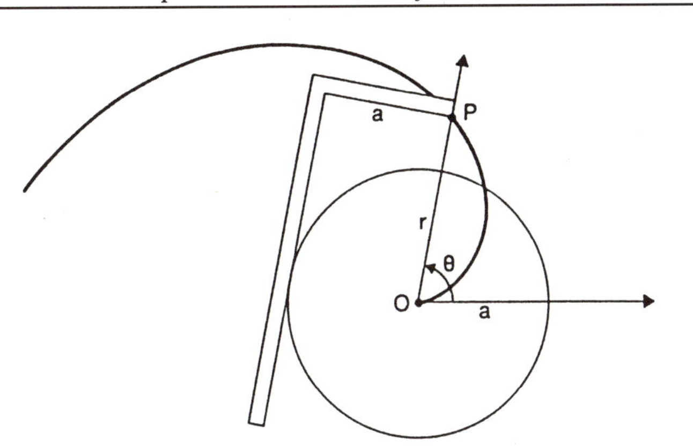

Figure 1.2. Archimedean spiral.

to which D'Arcy Wentworth Thompson's remarkable book, *On Growth and Form*, testifies eloquently.

The turning of a snail shell or the shell of a nautilus is a so-called spiral. The word *spiral* in its mathematical denotation refers to a plane curve described by a point P rotating about a fixed point O in the plane while simultaneously moving away from O.

There are many spirals. The spiral shown as the boldface curve in Fig. 1.2 was studied with particular interest by the great Greek mathematician, physicist, and engineer Archimedes (287–212 B.C.), and is now called an *Archimedean spiral.* This spiral is produced as the trace of a point P moving at constant velocity out along a half-line whose initial point is O, while simultaneously the half-line rotates at constant velocity in the plane. If the distance from the fixed point O to the moving point P is denoted by r, then an Archimedean spiral is determined by the equation $r = a \cdot \theta$, where θ is the angle described by the line segment OP measured from a half-line of

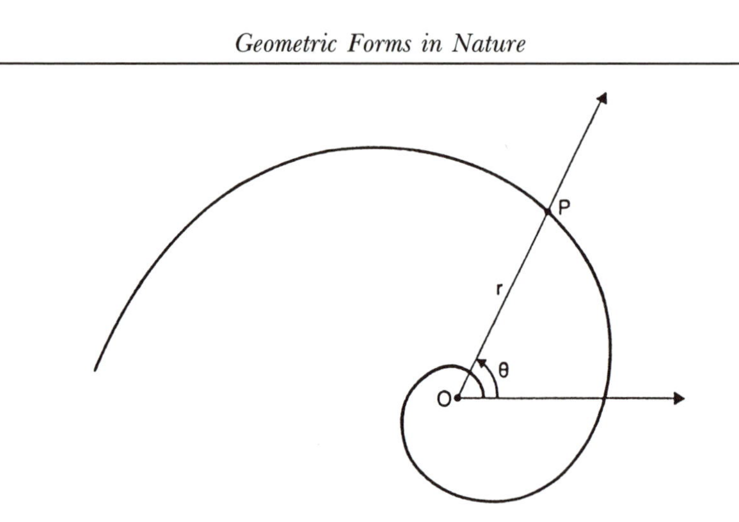

Figure 1.3. Logarithmic spiral.

fixed position in the plane beginning at point O, and a is a characteristic constant for the spiral. As shown in Fig. 1.2, an Archimedean spiral can be constructed by letting a right angle, whose shorter side has length a, roll along a fixed circle of radius a. It is thought that Archimedes tried to use this spiral to solve the classical problem of constructing with ruler and compass a square whose area equals some prescribed circle (*squaring the circle*), a task proved impossible more than 2,000 years later—one of the many consequences of the fact, demonstrated by the German mathematician Lindemann in 1882, that the number π is transcendental (no finite sum of integral multiples of powers of π is zero).

Now, however, we shift our focus to another spiral, namely, the logarithmic spiral shown in Fig. 1.3. This is the spiral that describes the turning of a snail shell, and it can be seen also as the curves formed by the seeds at the center of a sunflower.

A *logarithmic spiral* is produced as the trace of a point P moving out along a half-line with a velocity whose increment is proportional to the distance from its initial point O (exponential growth), while simultaneously the half-line rotates at constant velocity in the plane. When the magnitudes r and θ have the same meaning as in the case of an Archimedean spiral, then a logarithmic spiral is described by an equation of the form $r = a \cdot \exp(b \cdot \theta)$, where the values a and b are characteristic constants for the spiral, and exp represents the exponential function. There is no limitation to the constant b, but a must be positive. The equation can alternatively be written in the form $b \cdot \theta = \ln(r/a)$, where ln is the natural logarithm function.

To explain the arrangement of a snail shell, we need to consider somewhat the shape of a curve.

In general, as indicated in the previous examples, we can think of a plane curve as the curved trace of a point P moving in a given plane. At any moment during this motion, the point has a velocity vector that describes the instantaneous motion of the point (Fig. 1.4). The direction of the arrow indicates the direction of the instantaneous motion, while the arrow's length indicates the speed of the motion at the given moment in time. One way of visualizing the velocity vector is to imagine P as a tiny light being filmed during its motion. If we were to view the film in slow motion, P would appear to be moving along a broken path of straight lines (a polygonal path). At every point along this broken line, the forward-oriented edge is approximately a measure of the velocity vector at that point. We get the same impression seeing the curve drawn on a computer screen if the computer is programmed to replace the curve segment between two parameter values in the chosen step length with a line segment. The line, which contains the velocity vector to the curve at P, is called the *tangent* of the

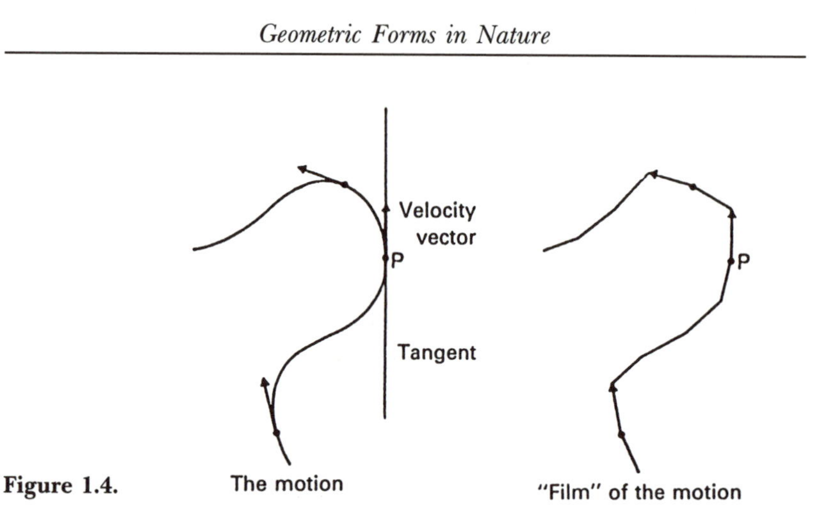

Figure 1.4.

curve at P. The tangent is the line that best approximates the curve in the neighborhood of the observed point.

Consider now a fixed point P_0 on the curve corresponding to the time t_0. For every point Q near P_0 on the curve, there is just one circle in the plane that passes through the points P_0 and Q and whose tangent line at P_0 contains the velocity vector for the curve at point P_0. As the nearby point Q approaches P_0, this circle approaches a limit circle (Fig. 1.5). The radius of the limit circle is called the *radius of curvature* for the curve under consideration at P_0, and the limit circle itself is called the *circle of curvature* of the curve at P_0. The circle of curvature is that circle that best approximates the curve in a neighborhood of the observed point. The *curvature* of the curve at point P_0 is the reciprocal of the radius of curvature. If in moving along the curve in the vicinity of P_0 one turns to the left, the curvature at P_0 is considered *positive*; if one turns to the right, the curvature is considered *negative*. Another way of

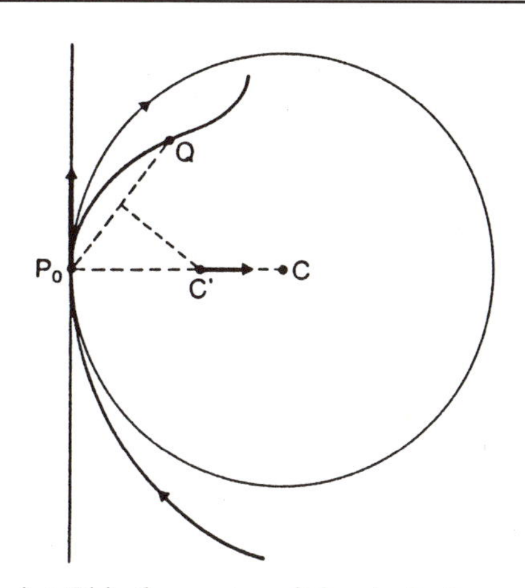

Figure 1.5. The point C′ is the center of the circle through P_0 and Q, whose tangent at P_0 contains the velocity vector for the curve at P_0. As Q approaches P_0, the point C′ approaches the center C of the circle of curvature of the curve at P_0.

saying this is that the curvature at P_0 is considered positive if the direction of travel around the circle of curvature induced by the direction of the curve's tangent is counterclockwise, and negative if it is clockwise. In Fig. 1.5, the curvature is negative at P_0. It may happen that the circles in question degenerate into lines; for example, if a segment of the curve around P_0 is a segment of a straight line. In this case, the curvature at P_0 is set to zero.

Consider now a logarithmic spiral, and look at the curve formed by the set of centers for the circles of curvature. It turns out that this curve is itself a logarithmic spiral that is congruent to the original curve. Now put the system of circles of curvature for the logarithmic spiral in such a way that the center of each single circle lies at the corresponding point on the curve and furthermore such that the circle is perpendicular to both the velocity vector at that point and the plane in which the logarithmic spiral is placed. Thereby we get a surface in space, and this surface is precisely a snail shell. The spiral in a snail shell is formed from the centers of the circles of curvature of the logarithmic spiral, and is thus itself a logarithmic spiral. How do the snails figure all this out? All we can say is, it is in the genes.

2. The Helix and the Twining Vine

The forms in nature often optimize circumstances. Indeed, one can say that nature is uncompromisingly efficient. Thus, architects and engineers can profit from observing how nature solves its problems. In this section, we cite a very simple example of this, namely, the twining growth of vines. In the course of our exploration, we shall discover the shortest path between two points on a surface.

Everyone knows that the shortest distance between two points A and B in a plane is a straight line segment connecting the points (Fig. 1.6).

Consider now a surface segment constructed of two planes that form a corner, as in Fig. 1.7. On each of the plane segments forming the corner in Fig. 1.7, a point is given, respectively, A and B. What is the shortest path on the surface segment between A and B? To solve this problem, we straighten out the corner to obtain the plane in Fig. 1.8. The two points A and B now correspond to the points A′ and B′. An arbitrary curve connecting A and B on the corner in Fig. 1.7 is transferred into a curve of equal length on the plane as

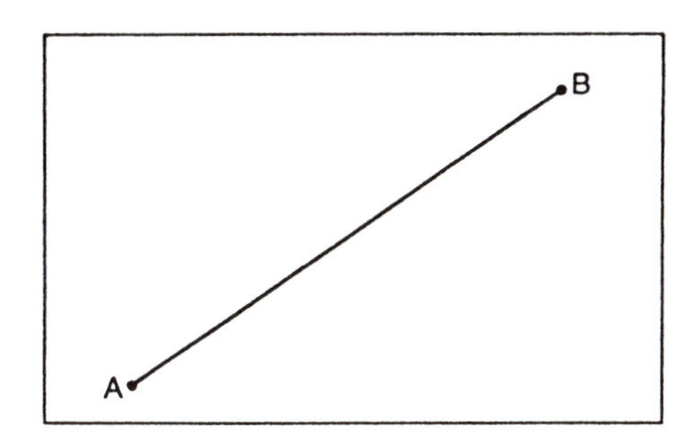

Figure 1.6.

displayed in Fig. 1.8. The shortest path on the corner between A and B thus corresponds to the shortest path in the plane between A′ and B′. The latter is of course the line segment connecting A′ and B′. The line segment from A′ to B′ intersects the line M′N′ corresponding to the intersection MN in the corner at an angle α. Thus, the solution to our problem is: The shortest path between A and B on the corner in Fig.

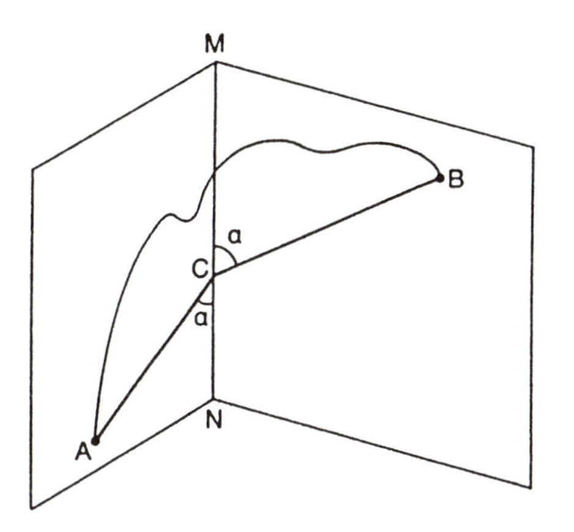

Figure 1.7.

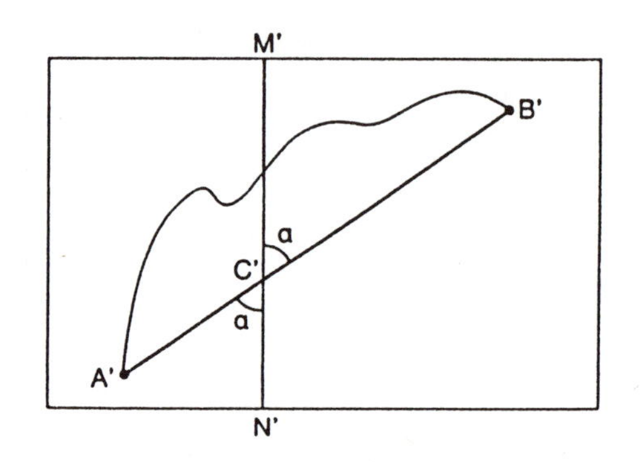

Figure 1.8.

1.7 is the broken line ACB, where C is the point on the edge of the corner MN determined such that MN forms an angle α with AC equal to the angle formed with CB.

Next, we proceed to consider two points A and B on the circular cylinder in Fig. 1.9. The cross-section of the cylinder is a horizontal circle, and the vertical lines that intersect the

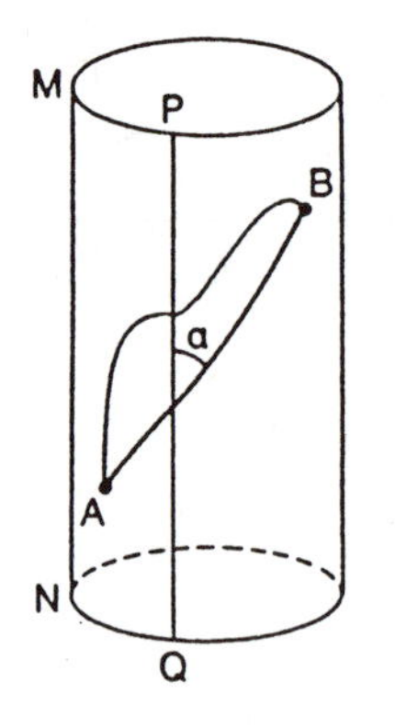

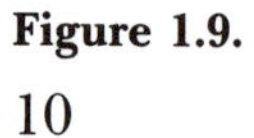
Figure 1.9.

circle generate the cylinder. What is the shortest path on the surface of the cylinder connecting the two points A and B?

To solve this problem, we cut the cylinder along its generating line MN, and spread it out into the plane rectangle of Fig. 1.10. The two points A and B are depicted here as A′ and B′, respectively. An arbitrary curve between A and B on the surface of the cylinder is transferred into a curve of equal length between A′ and B′ on the rectangle. The shortest path on the rectangle between A′ and B′ is the straight line segment that connects the points. This line segment can be characterized as that curve on the rectangle between A′ and B′ that intersects all vertical generating lines P′Q′ in the rectangle at a constant angle of α. When we roll it back onto the cylinder, we obtain the shortest path on the surface of the cylinder between A and B. Thus, the solution to our problem is: The shortest path on the surface of a cylinder between points A and B is that curve on the surface of the cylinder between A and B that intersects all generating lines PQ for the cylinder at a constant angle of α.

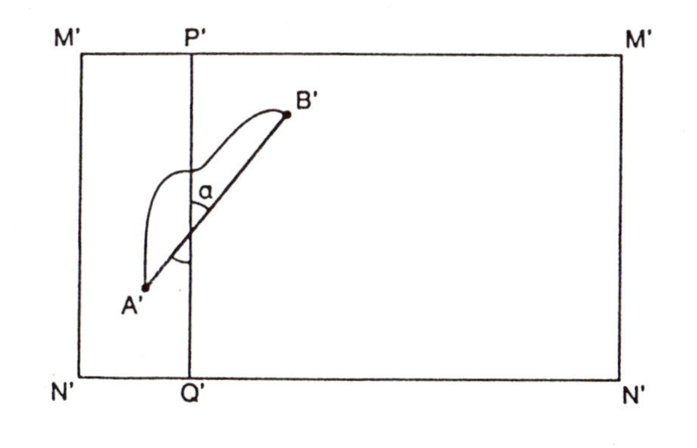

Figure 1.10.

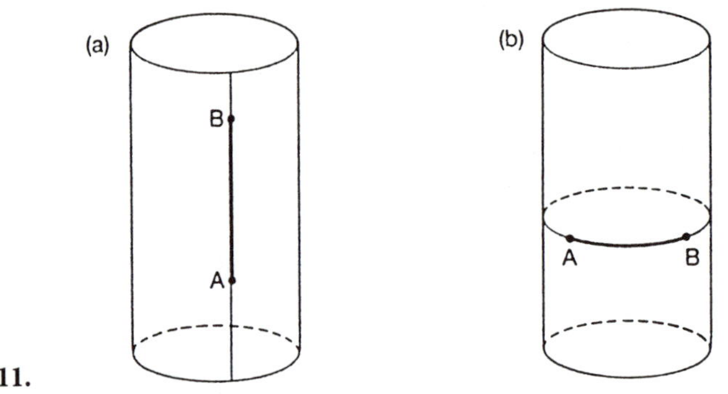

Figure 1.11.

In Fig. 1.11, we have two special situations, namely,

(a) A and B lie along the same generating line, and
(b) A and B lie along the same cross-section of the cylinder.

When A and B lie along the same generating line, the line segment AB on the generating line will be the shortest path from A to B. When A and B lie along the same cross section, the shorter of the two circular arcs from A to B is the shortest path on the surface of the cylinder from A to B.

Now consider a point moving on the surface of the cylinder. If the motion occurs along a generating line, we call it *vertical* motion; positive if upwards, negative if downwards. If the motion follows a circular track along a cross section of the surface of the cylinder, we call it a *rotation*; positive if counterclockwise (seen from above), negative if clockwise. If we have a point on the surface of the cylinder being drawn vertically upwards at a constant speed and drawn simultaneously around the cylinder along a circle (a cross section) at a constant speed,

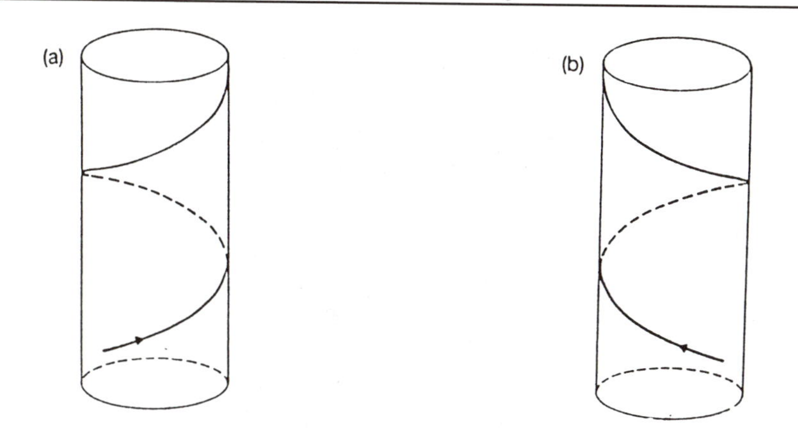

Figure 1.12. (a) Right-handed helix. (b) Left-handed helix.

the result will be that the point moves along a twisting curve that intersects all the generating lines at a constant angle. Such a curve is called a *helix*. The helix is called *right-handed* if the point moves with positive vertical motion together with positive rotation, as in Fig. 1.12a. If the point moves with positive vertical motion together with negative rotation, as in Fig. 1.12b, we call the helix *left-handed*.

Most twining plants (bindweed, for example) wrap themselves up around a cylindrical support (a stick) along a right-handed helix (botanists call them *levorotatory*), as in Fig. 1.13, while, for example, hops grow along a left-handed helix (botanists call them *dextrorotatory*), as in Fig. 1.14.

From our discussion of the shortest path on the surface of a cylinder, we can now derive the following result: *The shortest path connecting two points A and B on the surface of a cylinder is a helix.*

Figure 1.13. Bindweed. *Photo: Ebbe Sunesen.*

In this sense, nature can be said to be uncompromisingly efficient. In its upward course, a twining plant pursues the shortest possible path relative to the number of turnings.

The helix appears in many other contexts; for example, in the famous double-helix model of DNA molecules of Watson and Crick. (See, for example, *Scientific American*, Vol. 243, No. 1, July 1980, 100–113.)

3. The Geometry of Soap Films

Take a closed loop of wire and dip it in a suitable solution of soap—or glycerine—and a thin membrane of soap will be

Figure 1.14. Hops. *Photo: Erik Thomsen.*

stretched within the boundary curve of the closed wire loop. In Fig. 1.15, we show some examples. Such experiments were described in 1873 in a book by the Belgian physicist Joseph Plateau. Since the boundary curve can be arbitrarily complicated, it seems far from obvious that a soap film will always be formed, let alone a smooth membrane without points, edges, or other singularities. The problem of the existence of such membranes is known among mathematicians as *Plateau's problem,* and work on the solution to this problem has occupied many prominent mathematicians in this century. Thus, in 1936, Jesse Douglas was awarded one of the first two Fields Medals in mathematics for his proof that Plateau's problem always has at least one solution, provided singularities are

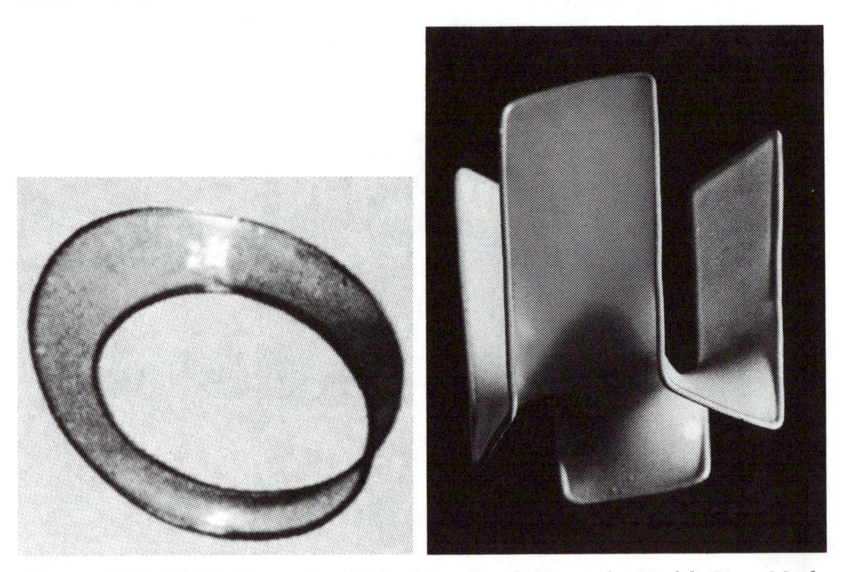

Figure 1.15. (a) Möbius strip. (b) Jenkins–Serrin's graph. *Model: Steen Markvorsen. Photo: Ove Broo Sørensen.*

allowed. The Fields Medal is given out every four years, most often to only two mathematicians. It is considered the highest recognition a mathematician can attain.

The study of Plateau's problem is part of a far more comprehensive problem, namely, of finding those surfaces with the least possible area relative to neighboring surfaces, when particular boundary conditions prescribed in advance are observed by the surfaces. A surface displaying minimal area locally is called a *minimal surface*. To be precise, this means that if we imagine the surface made of rubber, and if around an arbitrary point on the surface we place a small fixed closed curve, then the surface segment around the point bounded by

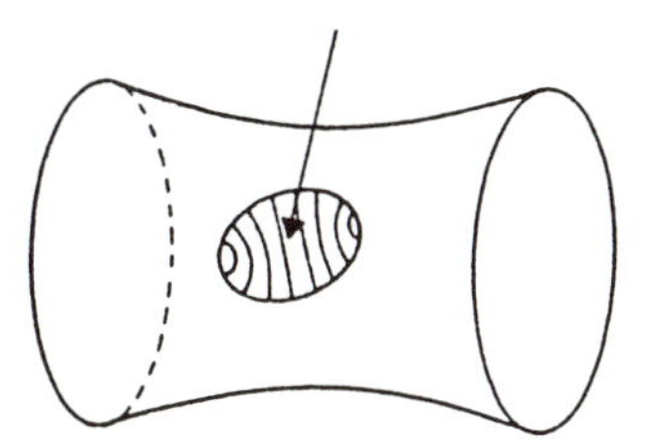

Figure 1.16. Minimal surface. The surface segment displays minimal area within the boundary curve. (See arrow.)

this curve will have minimal area among all surface segments we can produce by deforming the rubber membrane within the boundary curve. See Fig. 1.16, where we have shown a well-known minimal surface, the so-called *catenoid*, to be described shortly. Consider the plane curve formed in a vertical plane by a chain hanging between two points, affected by only the force of gravity; mathematicians call such a curve a *catenary*. The catenoid is that surface in space generated by rotating the catenary around an appropriate axis in the plane containing the catenary. In the 1740s, the astonishingly productive mathematician and physicist Leonhard Euler (1707–1783) discovered that the catenoid locally minimizes area; it is the only surface of revolution in space with this property.

The concept of a minimal surface can also be described using the notion of curvature, as in Section 1.1. Consider a smooth segment of the surface, where a positive normal direction has been chosen at every point. If through a point P on the surface we put a plane containing the normal and a given tangential direction to the surface at P, then a plane curve is cut in this normal plane to the surface—a so-called *normal section*. To every normal section we can ascribe a curvature, whose sign is determined by the rotational direction in

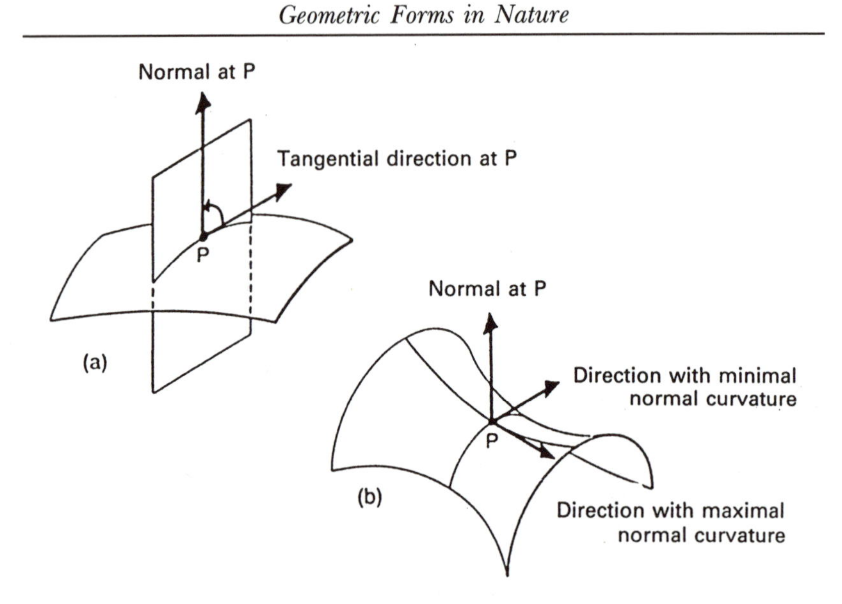

Figure 1.17. Normal section in a given tangential direction and the principal directions at P.

the sectional plane with the direction from the surface tangent to the surface normal as the positive rotational direction (counterclockwise), as in Fig. 1.17a. If we now rotate the normal section 360° around the surface normal, we obtain in the process a bunch of normal curvatures of the surface at the point P, of which one is maximal and one is minimal. These two curvatures are called the two *principal curvatures* of the surface at P. Those directions in which the principal curvatures are obtained are called the *principal directions* (Fig. 1.17b). The mean value of the two principal curvatures at P is called the *mean curvature* of the surface at P. It now holds that the surface is a minimal surface precisely when it displays zero mean curvature at all points. This formulation of the question of minimal surfaces was given by the French mathematician Meusnier in a paper written in 1776, and it has proven extremely fruitful. Meusnier himself knew that a *helicoid*, produced by drawing a line through every point on a helix orthogonally to the turning axis of the helix, has a mean curvature of zero at every point, and that this is the only ruled surface in space with this property. The helicoid is one of the surfaces in Fig. 1.20.

Many examples of minimal surfaces are found in nature. An oft-cited, though in this context perhaps somewhat dubious, nearly minimal construction is the honeycomb. (In the article, "What the bees know and what they do not know," published in the *Bulletin of the American Mathematical Society* **70** (1964), 468–481, L. Fejes Tóth has set forth how near the honeycomb is to being a minimal construction.)

Minimal surfaces offer elegant applications in architecture, as manifest in the works of the German architect Frei Otto, among others. Many of his elastic, lightweight constructions are transferred directly from models of minimal surfaces. This is true for example of his utilization of soap film geometry in the periodic minimal surfaces he used to build the German Pavilion at Montréal in 1967, and in the roof construction of the 1972 Olympic Stadium in Munich (Fig. 1.18). (See, e.g., *The Mathematical Intelligencer* **4**(4) (1982), p. 169, and Hildebrandt and Tromba, *Mathematics and Optimal Form*, Scientific American Library, W. H. Freeman and Co., 1985, pp. 136–137.)

The appeal of minimal surfaces for engineers and architects rests especially on the following: Minimal surfaces can be very pleasing aesthetically; minimal surfaces optimize the utilization of materials; minimal surfaces are strong, since loading in the

Figure 1.18. The Olympic stadium in Munich. *Photo*: *Jørgen Jensen*.

normal direction is counteracted by stretching forces in the material in at least one direction; minimal surfaces possess natural geometrical rigidity.

Within the last 10 years, significant advances have been made with respect to Plateau's problem. Thus, it has now been proved that an arbitrary closed curve in space can suspend a regular minimal surface without "holes," provided self-intersections of the surface are allowed. The solution need not be unique. It has still not been completely clarified under what conditions the solution is unique, or when there exists a solution without self-intersections.

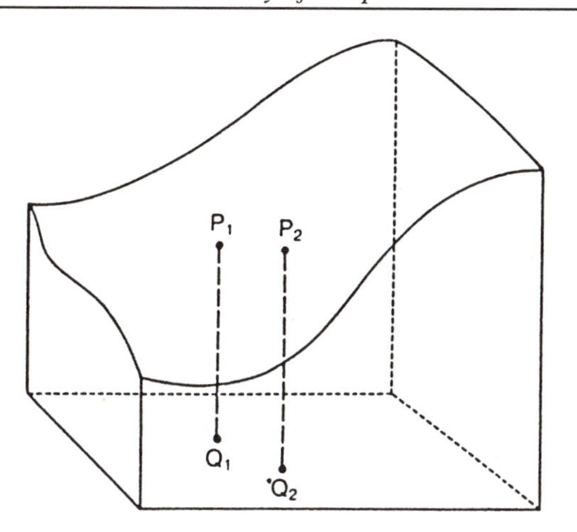

Figure 1.19. In a right-angled projection of the surface onto the plane, any pair of different points P_1 and P_2 on the surface are projected onto a pair of different points Q_1 and Q_2, respectively, in the plane. (The projection is true.)

Another interesting question relating to minimal surfaces, in particular their generalization to higher dimensions, concerns the so-called *Bernstein problem.* In a 1910 paper, Bernstein proved that a surface in space that permits a true right-angled projection onto a full plane in space (Fig. 1.19) is minimal precisely when it is itself plane. This raises a natural question (the Bernstein problem) in all dimensions n higher than 3 concerning the structure of a minimal, $(n - 1)$-dimensional geometrical object that permits a true right-angled projection onto an $(n - 1)$-dimensional hyperplane. With respect to this question, Bombieri, de Giorgi, and Miranda proved in 1969 that solutions to the Bernstein problem exist in dimension 9 that are not $(n - 1)$-dimensional hyperplanes. This proof was one of the achievements for which Bombieri was awarded the Fields Medal in 1974.

Since the middle of the 1980s, computer graphics have assumed a central position in the investigation of minimal surfaces. This began with a young Brazilian mathematician named Celsoe Costa, who in his doctoral dissertation described equations for an entirely new and heretofore unknown family of minimal surfaces in space. In their studies of computer graphics, two American mathematicians, David Hoffman and Bill Meeks, became convinced that the surfaces could be generated by removing three points from a sphere with an arbitrary number of handles g (a closed surface of genus g, a geometrical object to which we shall return in Chapter 2). Subsequently, they managed to prove by geometrical and analytical reasoning, as is proper in mathematics, that this really was the case. In Fig. 1.20, we show some computer pictures of minimal surfaces produced by James T. Hoffman. Included are some of the new types of minimal surfaces.

A corresponding, though slightly different, circle of ideas arises in connection with soap bubbles, which are, after all, inflated films of soap. Mathematically, we are dealing here with closed surfaces, where the mean curvature is constant over the whole surface. The geometry of soap bubbles is described in the fascinating book *Soap Bubbles*, by C. V. Boys, first published around the turn of the century, and reprinted in the Dover series in 1958. This book also offers practical advice on recipes for producing the best soap bubbles.

4. The Geometry of Tiled Surfaces

When we admire tiled surfaces, or more generally, the decoration of borders and plane areas with regular patterns that

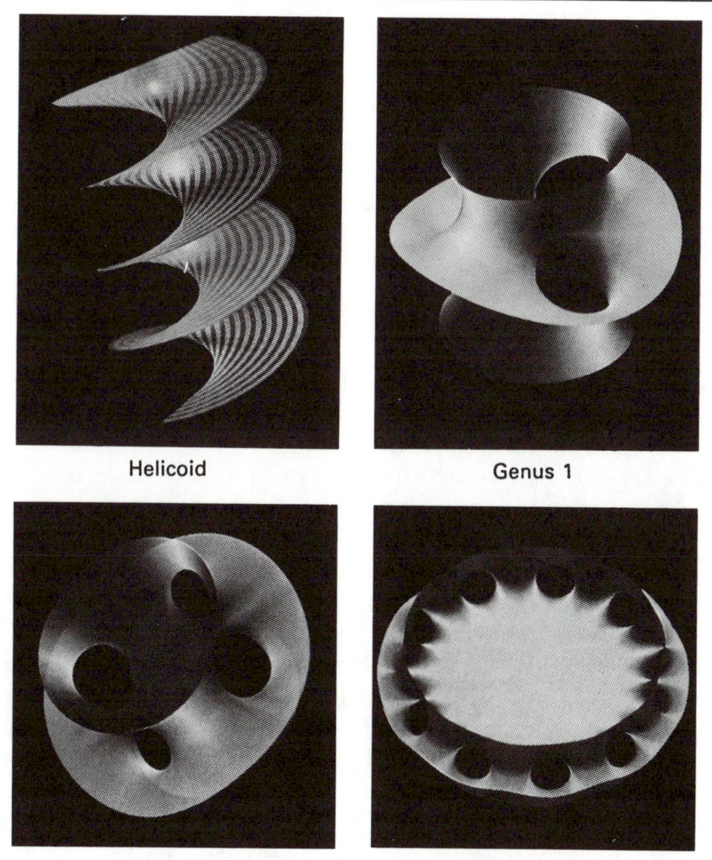

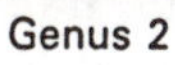

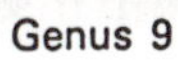

Figure 1.20. These illustrations are drawn with permission from David Hoffman, "The Computer-Aided Discovery of New Embedded Minimal Surfaces," *The Mathematical Intelligencer*, Vol. 9, No. 3, Springer-Verlag (1987), pp. 8–21. (*Thanks to David Hoffman and Springer-Verlag for permission to use the illustrations.*)

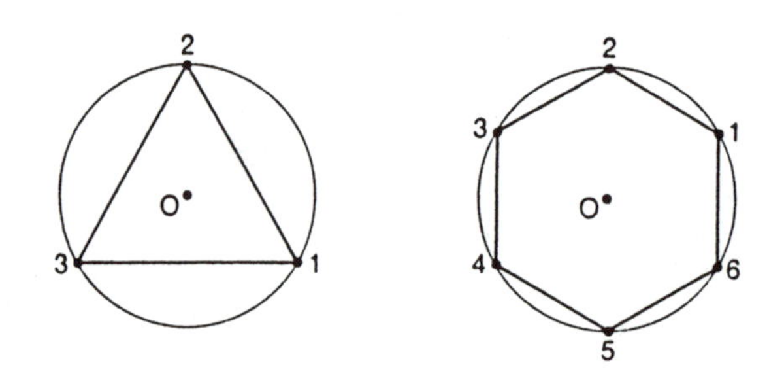

Figure 1.21.

repeat themselves periodically (*ornamentation*), the possibilities might appear endless. If we require regularity and symmetry, however, that is far from the case.

Let us see which possibilities exist if we wish to cover a level area with identical tiles in the shape of a regular n-sided polygon. A *regular n-sided polygon*, where $n = 3, 4, 5, \ldots,$ is a natural number, is constructed by dividing a circle into n equal parts and drawing the inscribed chords as in Fig. 1.21, where we have shown the regular triangle and hexagon.

Let us now ask: For which natural numbers $n \geq 3$ can we fill out the plane with identical regular n-sided polygons? As shown in Fig. 1.22 it is possible to do so for $n = 3$, 4, and 6. Is it also possible for $n = 5$? The answer is no, since the angle between two adjacent sides in a pentagon is 108°, and 360 cannot be divided evenly by 108. See Fig. 1.23, where we apply the fact that an inscribed angle measures half of the arc it spans. For an n-sided polygon, the corresponding angle is

$$\frac{1}{2} \cdot \frac{n-2}{n} \cdot 360° = \left(1 - \frac{2}{n}\right) \cdot 180°.$$

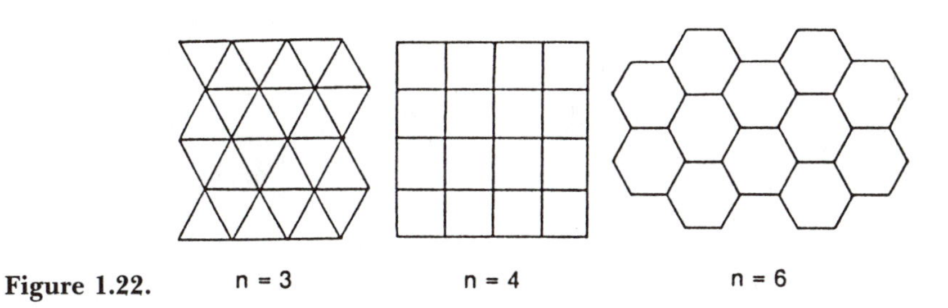

Figure 1.22.

Since $(1 - 2/n) \cdot 180 > 120$ for $n > 6$, and since every vertex in a covering of the plane with identical regular n-sided polygons must always be a vertex in at least three n-sided polygons, we have to exclude $n > 6$ as well. Thus, we have proved: *The only values of n for which one can fill out the plane with identical regular n-sided polygons are* $n = 3$, 4, and 6.

The patterns in Fig. 1.22 show familiar tile arrangements, of which the hexagon recently appears to have gained considerably in popularity. From the insect world we all know another common filling out of a surface with regular hexagons, namely a cross section in the comb of the honeybee (Fig. 1.24). It can be mathematically proved that the hexagonal elaboration of the honeycomb's cross section affords optimal utilization of the beeswax.

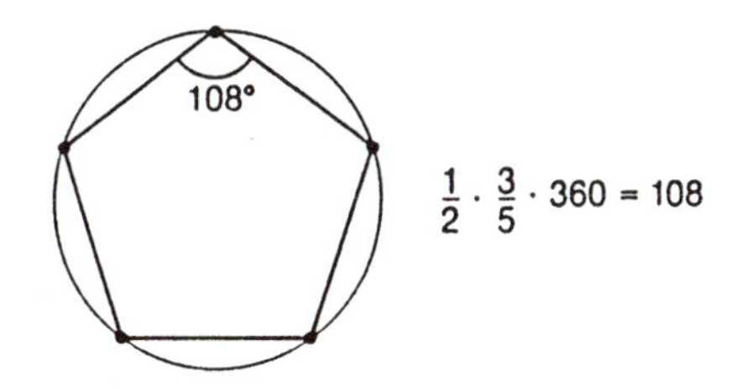

Figure 1.23.

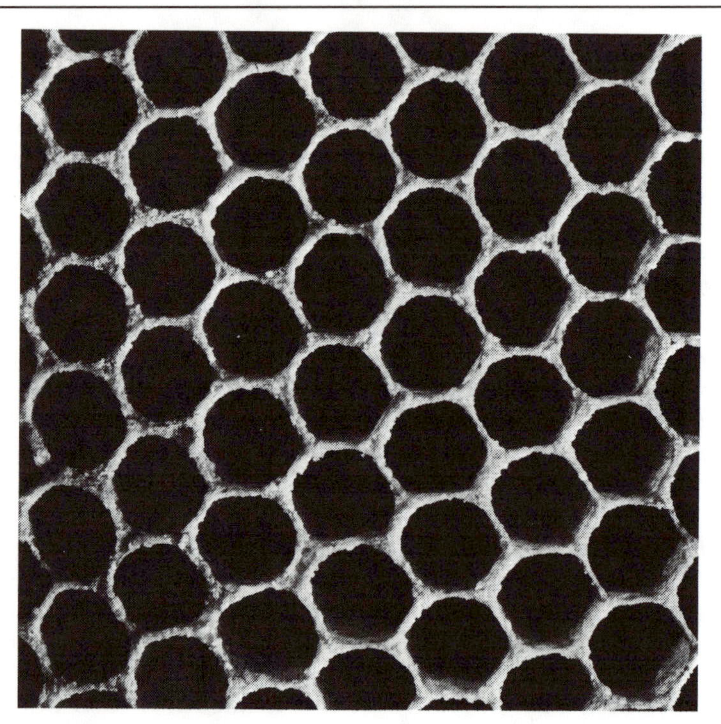

Figure 1.24. Comb of the honeybee. *Photo*: *Elvia Hansen*.

Looking beyond regular polygons to arbitrary shapes, it turns out there are just 17 different possible arrangements of symmetry if the shapes are to fit together by parallel displacement in two different directions; we call this *double periodic* parallel displacement. The degree of symmetry of a plane shape is measured by the set of mirror symmetries and rotational symmetries the shape permits. The regular triangle, for instance, displays three mirror symmetries and three rotational symmetries, since the zero rotation that maintains the shape, too, is counted as a rotational symmetry. The set of symmetries of a shape is an example of what mathematicians call a *group*. Precisely put, the foregoing means that only 17 different symmetry groups for plane shapes that display double periodicity can occur. This result was proved by George Pólya in 1924 in the paper, "Über die Analogie der Kristallsymmetrie in der Ebene," in Vol. 60 of *Zeitschrift für Kristallographie*, pp. 278–282. The 17 possibilities for a full symmetry group are represented in Fig. 1.25 by just those patterns that Pólya presented in his article. Many of these patterns were known already to the Egyptians, and they play a prominent role in the Arab decorative arts. Examples of these patterns can be found in the Alhambra, palace of the Moorish kings, built in the 14th century in Granada in southern Spain.

Things become more complicated if one discards the requirement of double periodicity, and takes into consideration only which figures can fill out the plane. The question is the second part of the 18th of the problems formulated by Hilbert at the International Congress of Mathematicians in Paris in 1900, where he summarized a number of fundamental questions in mathematics that remained unanswered at the time. The question is whether there exist asymmetrical figures that can fill out a plane, or more generally, the analogous question in higher dimensions, but where the filling out cannot be performed by moving the model figure around by a (discrete) group of Euclidean motions (a so-called *Bieberbach group*). The first example of such a figure was given by Heesch in 1935, and it is shown in Fig. 1.26a together with other examples of asymmetrical figures that can fill out a plane. It requires a little fiddling to convince oneself that the figure suggested by

Figure 1.25. Patterns corresponding to the 17 different symmetry groups of plane shapes.

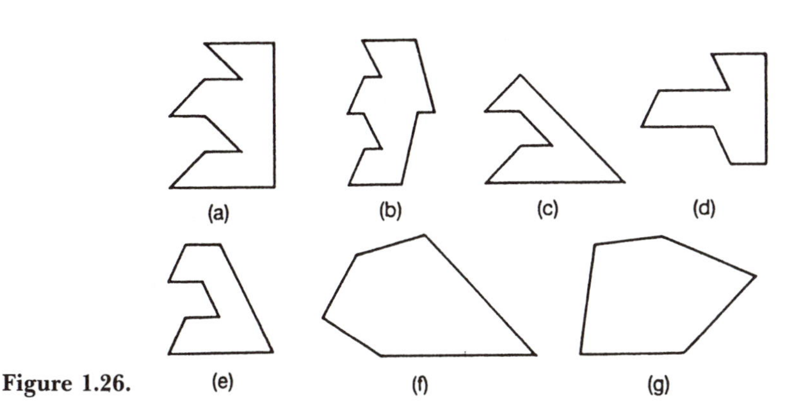

Figure 1.26.

Heesch can indeed fill out a plane, but it is not impossible and let this be a challenge to the reader. There is still considerable research activity going on in this area.

5. The Regular Polyhedra

In the plane, there exists a regular n-sided polygon corresponding to every natural number $n = 3, 4, 5, 6, \ldots$ The corresponding configurations in space are the regular polyhedra. Here, however, their number is strictly limited. As early as the time of the Pythagoreans, it was known that there exist only five regular polyhedra in space. We shall prove this in the present section.

By a *polyhedron*, we mean (the surface of) a three-dimensional body bounded by polygonal lateral faces. The polyhedron is called *convex* if the three-dimensional solid body together with any two arbitrarily chosen points on its surface contains the line segment connecting those points. A convex polyhedron, in which all the polygonal lateral faces are identical (congruent) regular polygons, and all dihedral angles between its lateral faces are equal, is called a *regular polyhedron.*

We shall need a theorem about convex polyhedra known as Euler's Theorem on Polyhedra, though it was known to Descartes in 1639, and through his unpublished manuscript also to Leibniz in 1675. In fact, we think it was known already to Archimedes. It is nevertheless Euler's name that has become attached to the theorem, since Euler provided the first published proof in a paper of 1750.

Euler's Theorem on Polyhedra says that if one considers the surface of a convex polyhedron (or the surface of a sphere divided into "bent" polygonal faces), and in this context count the following numbers,

v = number of vertices
e = number of edges
f = number of polygonal faces,

then it will always hold that $v - e + f = 2$.

We shall suggest an alternative proof, which according to tradition was employed by the Danish mathematician Johannes Hjelmslev (1873–1950). He ties the proof to Fig. 1.27. First, we position the polyhedron between two parallel horizontal planes, each of which touch the polyhedron at just one vertex. We can choose these planes so that the system of horizontal planes lying in between each includes at most one vertex of the polyhedron. Now let us imagine the polyhedron slowly filled with water, so that each of the interjacent planes forms the surface of the water at the respective point in the process of filling. During the filling, we count continuously the alternating sum $v - e + f$ among the vertices, edges, and faces that

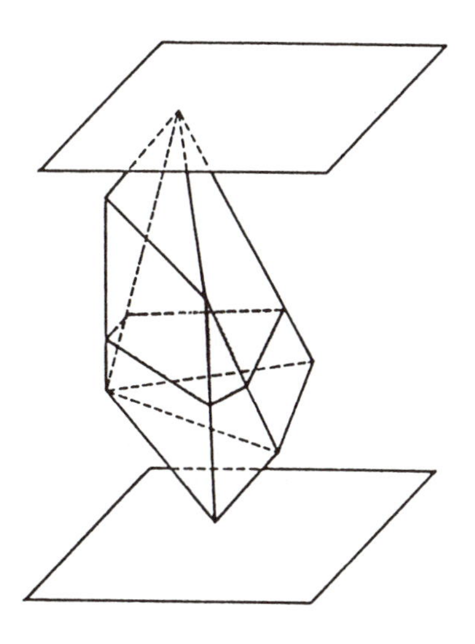

Figure 1.27.

have become wet at the water level we have reached. When the water level is zero, we are starting at the bottom with one vertex, that is, with $v - e + f = +1$. Each time we pass a new vertex as the water level rises (the filling still incomplete), the edges and faces around this vertex, which become wet before the next vertex does, will cancel each other in the alternating sum $v - e + f$, except for one edge which will be cancelled by the vertex. When the polyhedron has been filled to the top, we are left with the uppermost vertex. This vertex cannot be cancelled, and so adds a +1. Altogether, thus we obtain—*summa summarum*—that $v - e + f = 1 + 1 = 2$, which should be proved.

Many other proofs of Euler's Theorem on Polyhedra exist. One such proof, by Cauchy in 1811, is reproduced in my article, "Fra Geometri til Topologi," in *Nordisk Matematisk Tidsskrift*, Vol. 36, No. 2 (1988), 48–60.

To the surface of a convex polyhedron (or, similarly, to a sphere), Euler's Theorem on Polyhedra associates a number, namely, the number 2, which depends on the qualitative geometrical structure alone. The number is independent, for example, of the sphere's radius. One might say that here lies the root to that branch of geometry that in Euler's time was called *geometria situs*, or *analysis situs*, and in our day is called *topology*.

As mentioned, Euler's Theorem on Polyhedra can be used to prove that there exist just five types of regular polyhedra. For this purpose, consider a regular polyhedron constructed of regular n-sided polygons such that at every vertex the same number m of n-sided polygons meet. It must hold that $n \geq 3$ and $m \geq 3$. If the number of faces is f, the number of vertices will be $v = (f \cdot n)/m$ (there are n vertices on each lateral face, and every vertex is included in m lateral faces), and the number of edges $e = (n \cdot f)/2$ (each edge lies along two lateral surfaces). Thus, Euler's Theorem on Polyhedra shows that

$$2 = v - e + f = \frac{f \cdot n}{m} - \frac{n \cdot f}{2} + f = f \cdot \left(\frac{n}{m} - \frac{n}{2} + 1\right).$$

Table 1.1.

n	m	f	Regular polyhedron
3	3	4	Tetrahedron
3	4	8	Octahedron
3	5	20	Icosahedron
4	3	6	Hexahedron
5	3	12	Dodecahedron

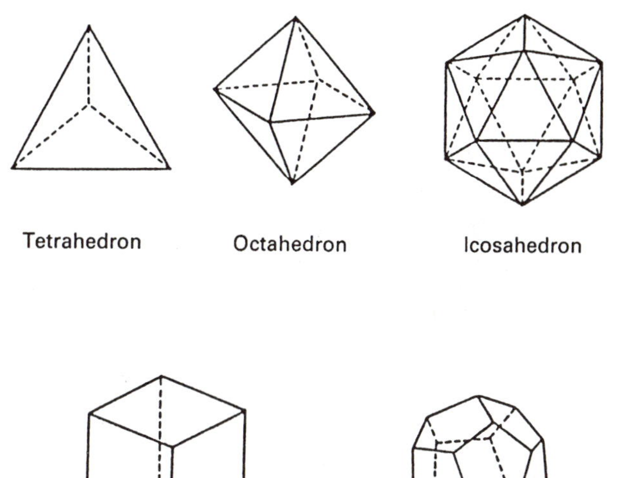

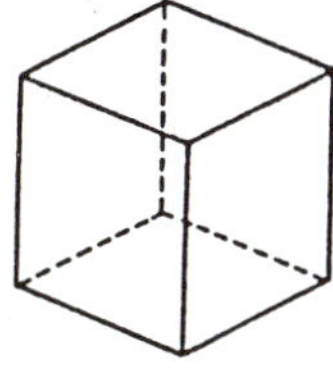

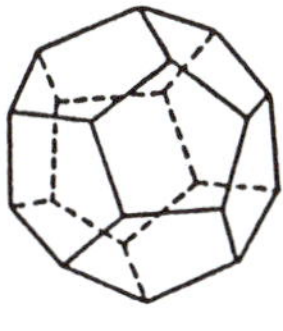

Figure 1.28.

Dodecahedron

Since $f > 0$, it must hold that $n/m - n/2 + 1 > 0$, or equivalently,

$$n \cdot \left(\frac{1}{2} - \frac{1}{m}\right) < 1.$$

Since $m \geq 3$, then $\frac{1}{2} - 1/m \geq \frac{1}{2} - \frac{1}{3} = \frac{1}{6}$, and therefore, $n \leq 5$. For n, therefore, the only possibilities are $n = 3, 4, 5$. Since $n \cdot (\frac{1}{2} - 1/m) < 1$, this yields corresponding limitations for m. In Table 1.1, we have presented the possibilities for matching pairs (n, m), together with the corresponding number of faces,

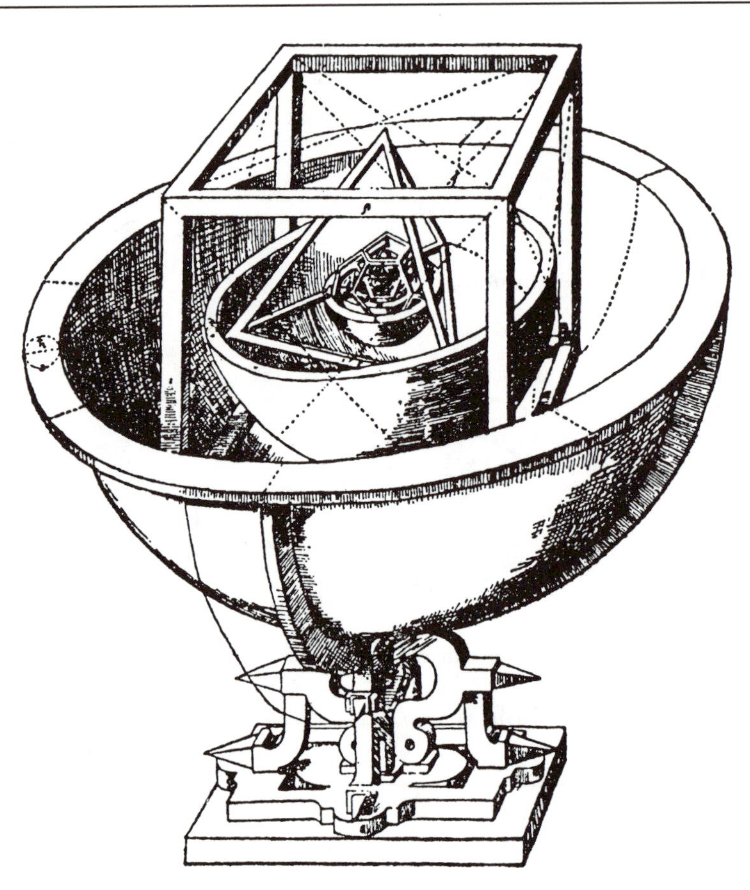

Figure 1.29. Kepler's model of the universe of 1595.

f, of a possible regular polyhedron, calculated from the formula $2 = f \cdot (n/m - n/2 + 1)$. It turns out that all the possibilities can be realized, and we give the names of the respective regular polyhedra: tetrahedron, hexahedron, octahedron, dodecahedron, and icosahedron.

The very classification of the regular polyhedra was known in antiquity to the Pythagoreans, who considered the construction of the dodecahedron as one of their greatest accomplishments. They play a prominent role in Plato's philosophy, and are known also as the *Platonic bodies*. In the dialogue, *The Timaeus*, Plato associates the tetrahedron, the octahedron, the hexahedron, and the icosahedron with the four elements, fire, air, earth, and water, respectively, while in the dodecahedron, he sees an image of the universe itself. The five Platonic bodies are pictured in Fig. 1.28.

Plato, a pupil of Socrates, lived from 427 to 347 B.C. Through the centuries, the Platonic bodies have retained their fascinating and alluring effect. Thus, in 1595, long before he discovered his three laws of planetary motion, Kepler attempted to reduce the distances in the planetary system to regular polyhedra inscribing and circumscribed by spheres. In Fig. 1.29, we show Kepler's construction. Kepler was convinced that with this construction he had thrust deep into the mystery of the creation and the order of the universe. The six spheres correspond to the six planets, Saturn, Jupiter, Mars, Earth, Venus, and Mercury, which in that order are separated by the hexahedron, the tetrahedron, the dodecahedron, the octahedron, and the icosahedron. Of course, Kepler had no knowledge of the three outer planets, Uranus, Neptune, and Pluto, discovered in 1781, 1846, and 1930, respectively.

2 THE TOPOLOGY OF SURFACES

It is to Leibniz's credit to have pointed out in 1679 the need for studying the qualitative characteristics of geometrical objects. He called this study *analysis situs* or *geometria situs*. In our day, these names have yielded to the term *topology*, a designation introduced by Listing in a paper in 1847. Although Euler's Theorem on Polyhedra from 1750 is now considered the first true result in topology, by far the major evolution of the field has taken place during the 20th century, when topology has been dubbed the queen of mathematics by the virtually universal mathematician, the Frenchman J. Dieudonné.

In topology, as noted, we examine the qualitative characteristics of a geometrical object: Is it connected? Does it have edges? Holes? The quantitative dimensions of an object retreat into the background. To a topologist, a soccer ball (a sphere) is equivalent to a football (an "ellipsoid"), while these objects differ from an inner tube (a torus), because no matter how vigorously you may pump the inner tube, you cannot eliminate the hole.

In this chapter, we shall work with the classification of closed surfaces such as the sphere and the torus cited previously. This topic has the advantage that we can develop a pretty good intuition about surfaces that is altogether necessary as a ballast when the topologists soar into the realms of higher dimensions. At the same time, the topic represents a milestone in the history of topology, where for the first time non-trivial geometrical objects were successfully classified solely on the basis of their topological characteristics. Even though the theorem including a complete proof of it has been known since at least the 1920s, it continues to fascinate, and it offers a valuable insight into how topologists think and work.

Topological concepts penetrate mathematics throughout, and as we shall see in Chapter 5, there are applications in theoretical physics as well.

1. Some Familiar Surfaces

We encounter surfaces everywhere—in landscapes, as the shapes of animals and other natural objects, as well as in works of art. Think of the sculptures of Henry Moore. In daily life we meet surfaces in furniture, lamps, vases, and countless other things. From an early age, we are thoroughly familiar with surfaces like those in Fig. 2.1.

Further along the road, some of us discover more arcane surfaces—surfaces that cannot be oriented. The classic example is the Möbius strip, first studied by the German mathematician A. F. Möbius (1790–1868). The *Möbius strip* is obtained by taking a rectangular strip and identifying (sewing or gluing together) a pair of opposite ends, after first giving the strip a half twist (Fig. 2.2).

It is well known that if we cut a Möbius strip lengthwise down the middle, the resulting surface is connected (Fig. 2.3).

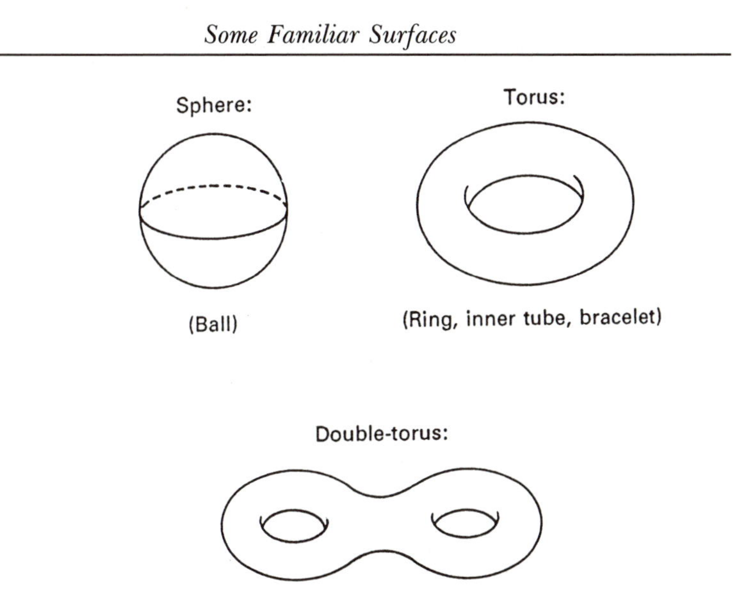

Chain:

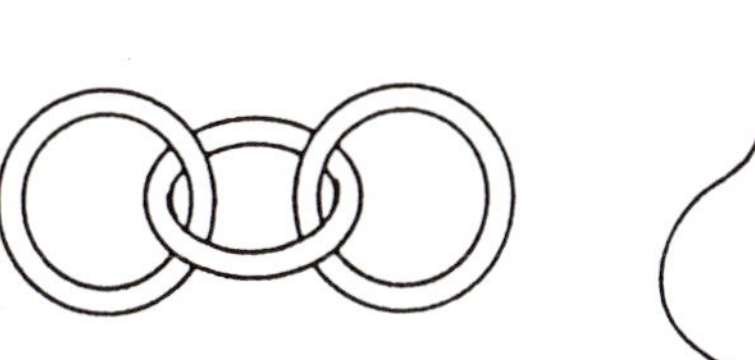

Vase:

Figure 2.1.

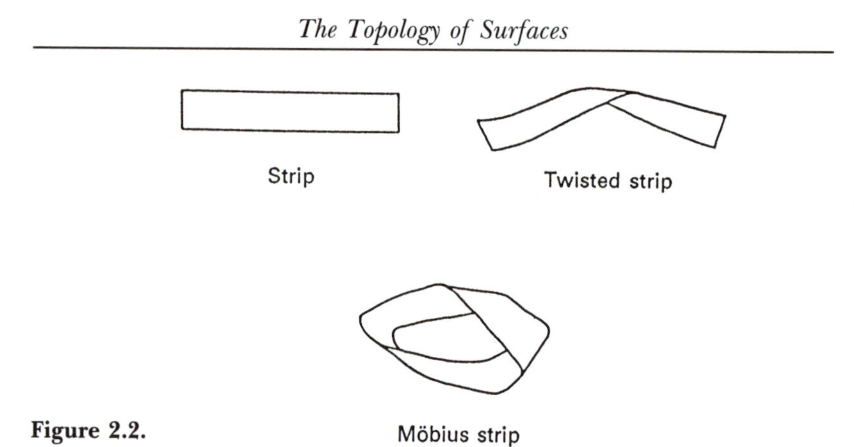

Figure 2.2.

It makes an amusing experiment to perform this cutting operation for a group of the uninitiated. Try cutting strips with more twists, too. In Fig. 2.3, we have marked the edge of the cut so we know how to sew it together to return to the Möbius strip.

Since the cut-up Möbius strip is a closed strip with two full twists, cutting the strip transversely, untwisting it, and then sewing it together again along the line we have just cut forms a surface equivalent to the cylinder in Fig. 2.4.

If we now identify a single pair of opposite points along the edge and pinch them together, we obtain the surface in

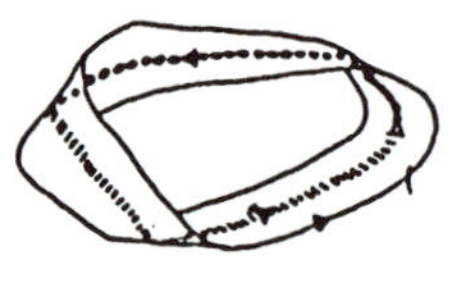

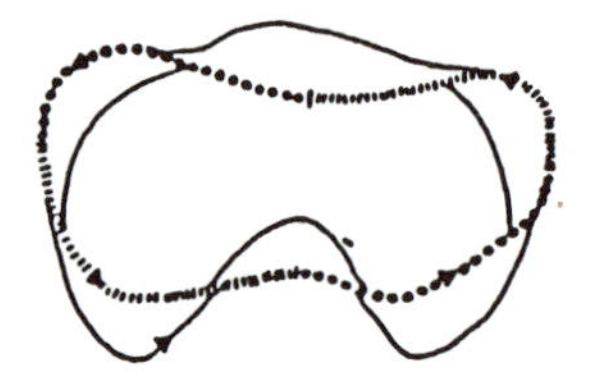

Figure 2.3.

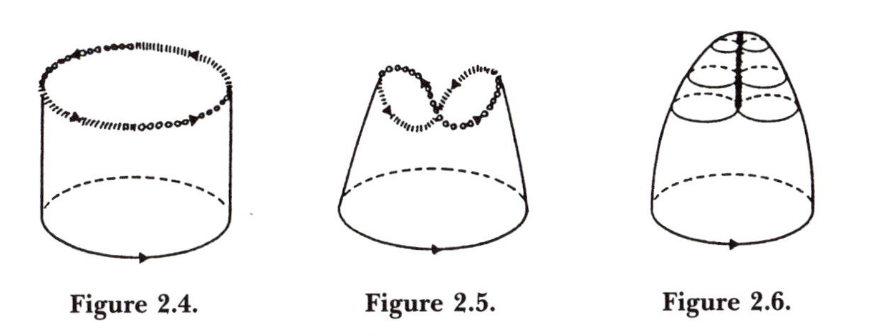

Figure 2.4. **Figure 2.5.** **Figure 2.6.**

Fig. 2.5. Purely abstractly, now we can sew it together in accordance with the indicated identifications. In normal three-dimensional space, though, this entails self-intersections. By sewing the surface together, we obtain the surface in Fig. 2.6. This surface is called a *crosscap*. The sewing process we have used is for obvious reasons called *diagonal sewing*.

Since we have sewn the surface together as prescribed by the identifications, we have in fact proved the following:

Proposition 2.1. *A Möbius strip is (topologically) equivalent to a crosscap.*

To a mathematician, a "proposition" is a statement less exacting than a "theorem," tied often to a noteworthy result, whose proof is obvious.

2. The Projective Plane and the Klein Bottle

In what follows, as a model of a sphere, or *2-sphere*, which we denote by S^2, we shall use the unit sphere with a distance 1 from a fixed initial point (*origin*) in space. The number 2 reminds us that when we look at a small segment of the sphere,

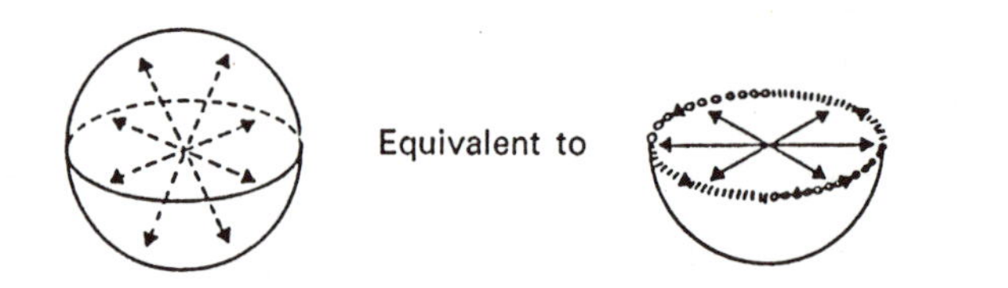

Figure 2.7.

we can describe it by a chart in which the points are determined by two coordinates. We say that the sphere has dimension 2. By the *antipodal point* to a point on the sphere, we mean the diametrically opposite point on the sphere.

We shall now consider two somewhat less familiar surfaces.

The projective plane, which we denote by $\mathbb{RP}^2$, is constructed by identifying every point on the sphere S^2 with its antipodal point, or equivalently, by identifying every point on the edge of a hemisphere with its antipodal point (Fig. 2.7).

In accordance with the definition, $\mathbb{RP}^2$, as is also shown in Fig. 2.7, can just as well be constructed by sewing the edge of a hemisphere diagonally. In ordinary space, this is again possible only if we permit self-intersections, as shown in Fig. 2.8. It follows from this that $\mathbb{RP}^2$ results from sewing a crosscap,

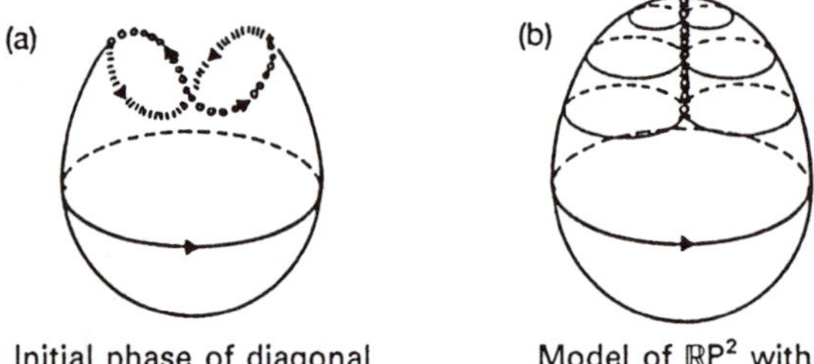

Figure 2.8. (a) Initial phase of diagonal sewing of the edge. (b) Model of $\mathbb{RP}^2$ with self-intersection.

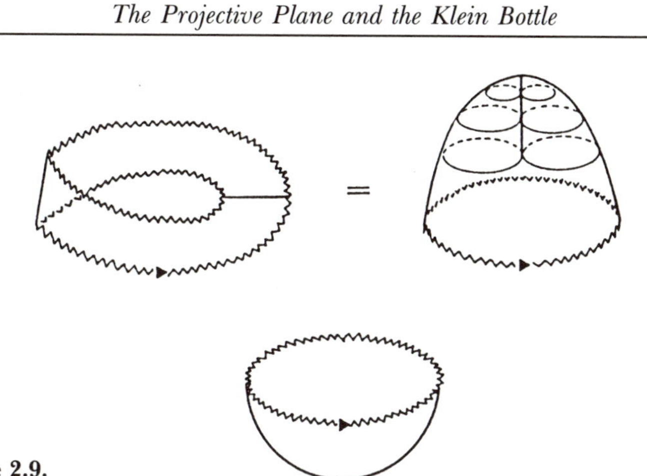

Figure 2.9.

or equivalently, in accordance with Proposition 1, a Möbius strip, on the edge of a hemisphere. See Fig. 2.9.

We state both definition and observations in the following:

Proposition 2.2. *The projective plane* $\mathbb{R}P^2$ *can be obtained by cutting a (curved) circular section out of the sphere and then performing one of the following equivalent operations:*

(1) *Sewing the edge diagonally.*
(2) *Sewing a crosscap lengthwise along the edge.*
(3) *Sewing a Möbius strip lengthwise along the edge.*

Another interesting surface was discovered by the German mathematician Felix Klein (1849–1925). This surface is now known as the *Klein bottle*. Abstractly speaking, we can define the Klein bottle, designated K^2, by giving the edge circles of a cylinder opposite rotational directions, and then identifying in agreement with these rotational directions. Once again, this

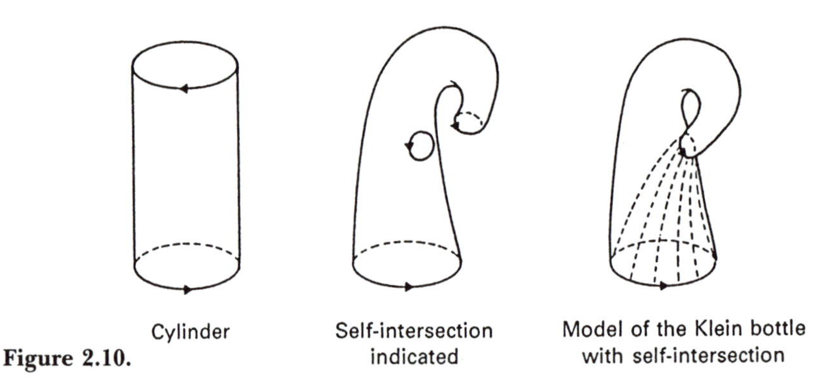

Figure 2.10.

surface can be realized in space only if we permit self-intersections. We undertake the identification by bending the cylinder around, sticking it through itself, and then sewing the edge circles together. The entire process is shown in Fig. 2.10.

If we imagine that three-dimensional space is supplemented with a fourth dimension, the self-intersection can be removed by lifting in this extra direction, in the same way as the self-intersection in a figure eight in the plane can be removed in space utilizing the third dimension by lifting perpendicularly from the plane, as shown in Fig. 2.11.

The series of pictures in Fig. 2.12 shows that the Klein bottle can be divided into two Möbius strips, and thus conversely can

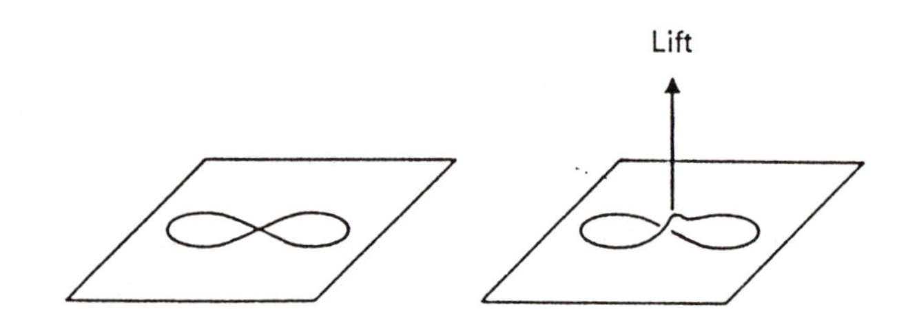

Figure 2.11.

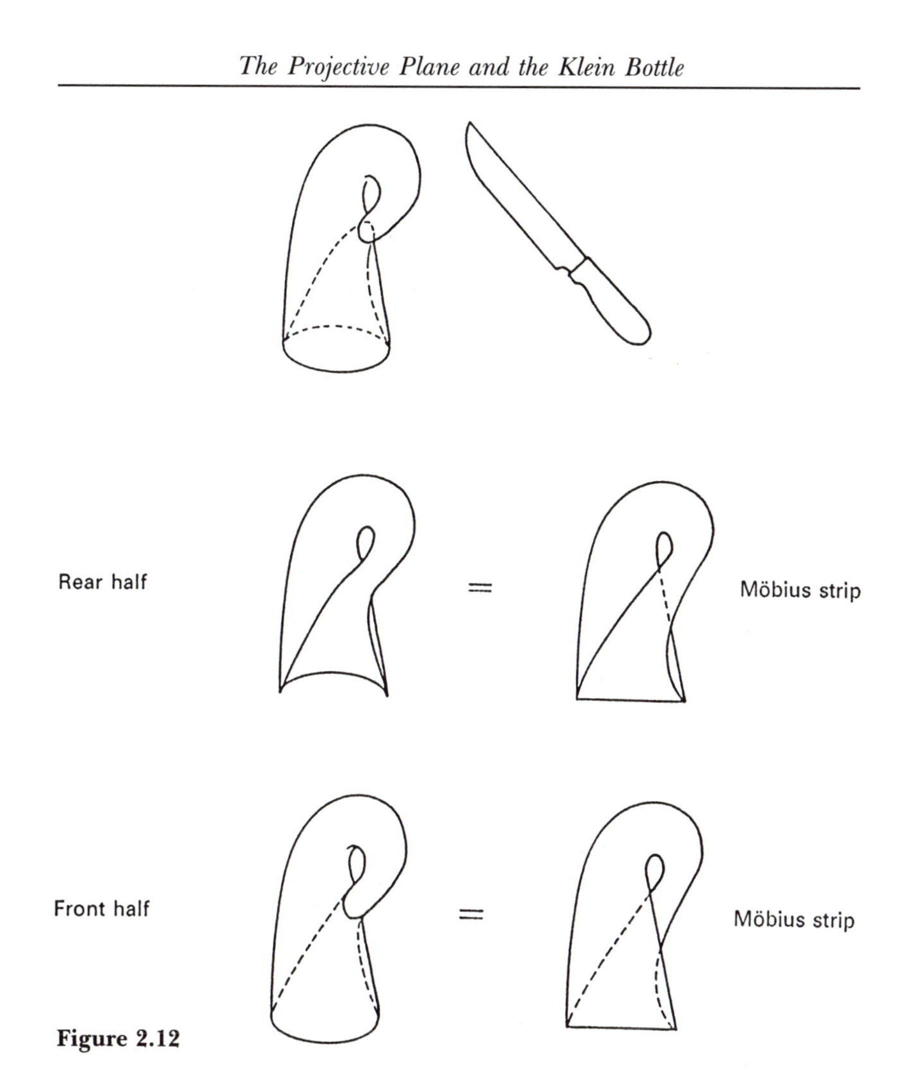

Figure 2.12

be constructed by sewing two Möbius strips together along their edges.

Once more exploiting the equivalence between the Möbius strip and the crosscap, we obtain the following:

Proposition 2.3. *The Klein bottle* K^2 *can be obtained by cutting two (curved) circular sections out of a sphere and then performing one of the following equivalent operations:*

(1) *Sewing diagonally along both circular edges.*
(2) *Sewing a crosscap lengthwise along each circular edge.*
(3) *Sewing a Möbius strip lengthwise along each circular edge.*

Let us pause here to note that we still have not defined the concept of a surface. We have deliberately not done so in the hope of demonstrating that we have such a well-developed intuition with respect to the concept of surfaces that we can work with them quite freely in the total absence of a precise definition. In the following section, we shall nevertheless try to state with greater precision what sort of object a surface basically is.

3. What Is a Closed Surface?

Intuitively, a *closed surface* is a geometrical object M, which resembles the plane locally, consists of a single piece (it is connected), and has no edges (the object is *closed*). Of the surfaces we considered in Section 2.1, we notice that only the chain is not connected, and that only the vase and the Möbius strip have edges.

To understand what is meant by saying that a geometrical object resembles the plane locally, one might think of how a

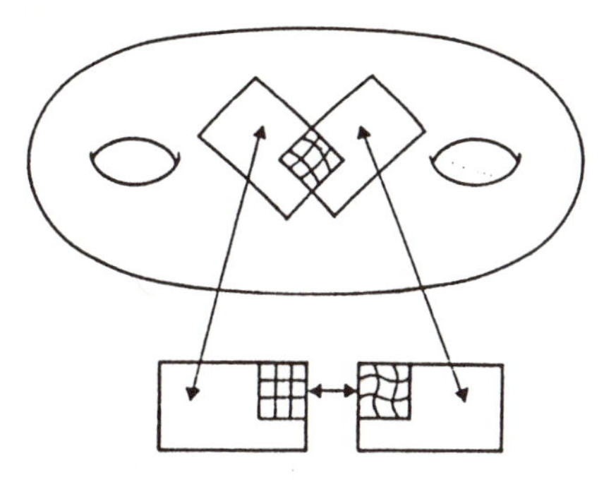

Figure 2.13.

world atlas represents the globe of the Earth. Each of the plane charts conveys the *local* information about its respective bit of the globe, while the atlas as a whole comprising all its local charts gives the *global* picture of the spherical Earth. In topology, we adopt the terminology of this illustration directly and say that a geometrical object resembles the plane locally if it can be described by an atlas of plane charts, where the local information agrees within overlapping charts. One can see at once that all the surfaces we have described so far resemble the plane locally in this sense. We say also that they are *two-dimensional manifolds*, as there are natural generalizations to higher dimensions. In Fig. 2.13, we have delineated two locally overlapping charts on a closed surface, and have indicated the charts' identification with plane areas.

If the local exchange of information among local charts that overlap one another occurs smoothly in the sense that smooth curves in one of the plane areas, corresponding to the common overlap of the charts on the surface, are transferred as smooth curves in the other area, as in Fig. 2.13, then the surface is said to be *smooth*. This is shown also by the fact that the surface naturally has a tangent plane at each of its points.

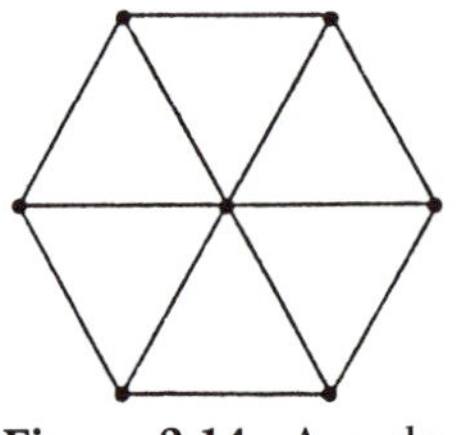

Figure 2.14. A cycle.

If the local exchange of information is linear in the sense that polygonal paths are transferred as polygonal paths, then the surface is said to be *piecewise linear*. The requirement that a surface must locally resemble the plane implies automatically—possibly after a shift of charts—that the charts overlap smoothly or piecewise linearly. For higher-dimension manifolds—a subject to which we shall return—this is far from always being the case.

In fact, a fundamental result proved by Rado in 1925 states that any closed surface (two-dimensional manifold) M can be *triangulated*. By this, we mean that M can be divided into a finite number of vertices, (curved) edges, and triangles so that

(1) If two triangles have anything in common, they have exactly one vertex or one edge in common.

(2) Each edge is an edge in exactly two triangles.

(3) The collection of triangles meeting in a common vertex can be arranged in a closed cycle, as in Fig. 2.14.

Triangulations of the sphere can be achieved by inflating a tetrahedron, octahedron, or an icosahedron. We shall not comment further here on triangulability, but shall be content to suggest triangulations of yet two important basic forms among closed surfaces, namely, the torus T^2 and the projective plane $\mathbb{RP}^2$.

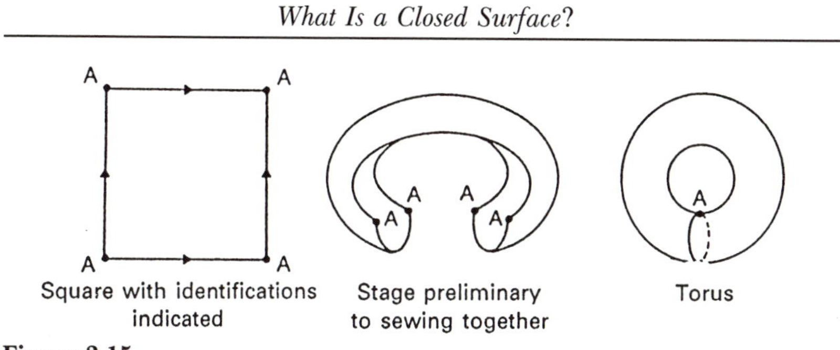

Figure 2.15.

The *torus* can be constructed by identifying (sewing together) the pairs of opposite sides of a square, as shown in Fig. 2.15.

A triangulation of the torus, containing the smallest possible number of vertices (7), edges (21), and triangles (14), is given in Fig. 2.16. Another triangulation of the torus is obtained by dividing all the quadrangles in the division of the torus in Fig. 2.17 by diagonals.

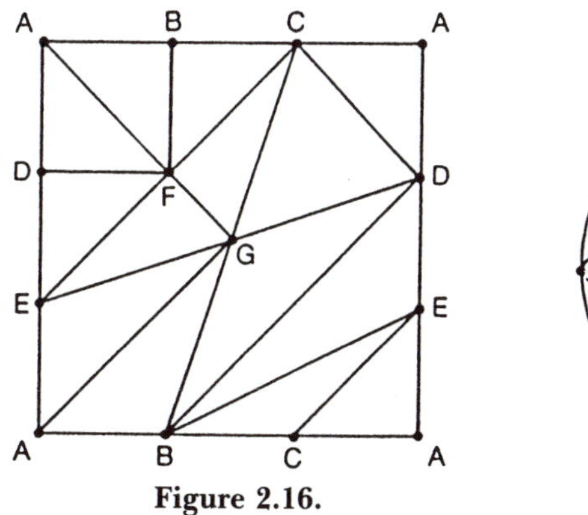

Figure 2.16.

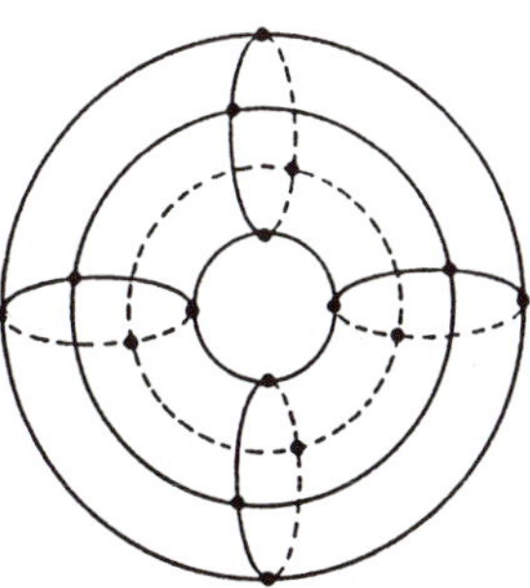

Figure 2.17.

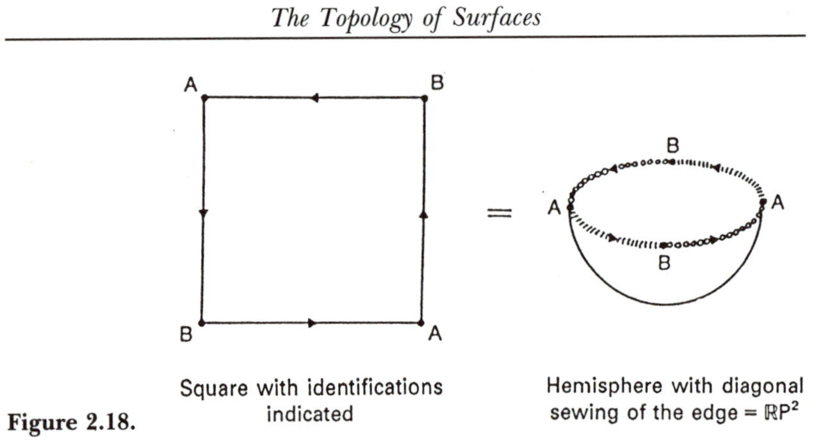

Figure 2.18.

The projective plane $\mathbb{R}P^2$ can be constructed by identifying in a square each pair of opposite sides with opposite orientations, as shown in Fig. 2.18.

A triangulation of $\mathbb{R}P^2$ containing the smallest possible number of vertices (6), edges (15), and triangles (10) is shown in Fig. 2.19.

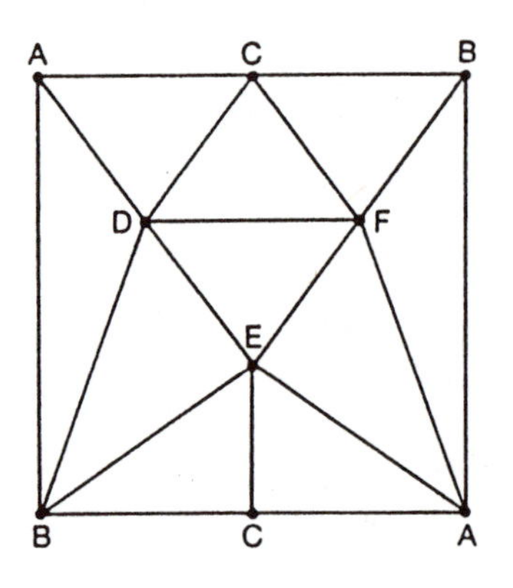

Figure 2.19.

4. Orientable and Non-Orientable Surfaces

An ant walking around a Möbius strip can reach any point on the surface without gnawing its way to the other side or crawling over the edge. It cannot do the same on a cylindrical surface. Try putting the ant on the two surfaces as Escher has done it on the Möbius strip in Fig. 2.20. The Möbius strip is an example of a non-orientable surface. There are many jokes about the Möbius strip—the one about the painter given the task of painting one side and stripping the paint off the other, for example. The poor man exhausted himself in the attempt.

An ant crawling around on the projective plane will also be able to reach every point on the surface, since the projective plane contains a Möbius strip, to which the ant can crawl and "turn." A sphere, on the other hand, contains no strip on which to "turn," so here an ant has to be either on the outside of the sphere or on its inside.

These intuitive reflections are crystallized in the concept of *orientation*. A closed surface is said to be *orientable* if it contains no Möbius strip. If a closed surface contains a Möbius strip, it is said to be *non-orientable*.

Of the closed surfaces we have encountered so far, the sphere, the torus, and the double torus are orientable surfaces, while the projective plane and the Klein bottle are non-orientable surfaces.

On a non-orientable surface, one can find a closed curve that runs in a strip equivalent to a Möbius strip, along which a chosen direction of travel (orientation) in a "plane" neighborhood of a fixed point on the curve will be reversed by transport around the closed curve.

Sometimes it is said that non-orientable surfaces have only one side, or that they have no inner side. On the Möbius strip,

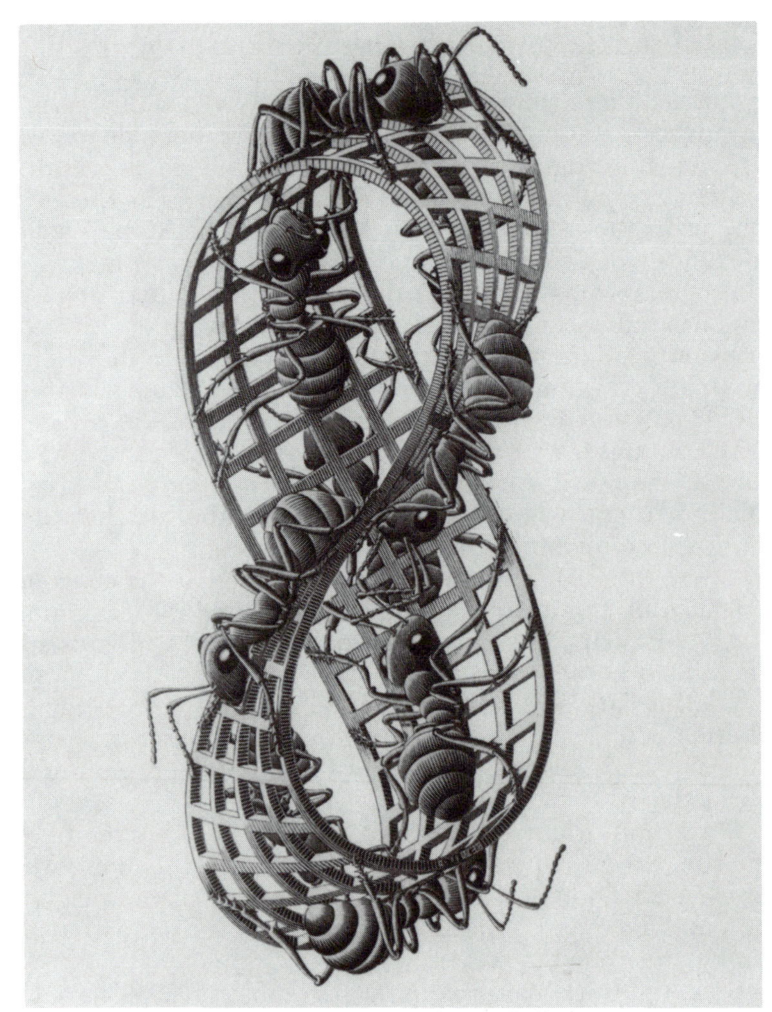

Figure 2.20. Möbius strip by M. C. Escher.

such notions make sense. On the Klein bottle, they do not. If one thinks of the model of a Klein bottle with its self-intersection in three-dimensional space, an ant crawling around on the surface cannot reach everywhere, since it will be blocked at the self-intersection. As discussed in Section 2.2, we overcome the self-intersection by ascending into four-dimensional space; but just as the circle cannot be said to have an inside and an outside in three-dimensional space, the Klein bottle cannot be said to have an inside and an outside in four-dimensional space.

5. Connected Sum of Closed Surfaces

Let M_1 and M_2 be two closed surfaces. Out of each of the surfaces, a (curved) circular disc or the interior of a triangle in a triangulation is cut. We obtain thereby two surfaces with edges, the edges in each case being identifiable with a circle or the edges of a triangle. If we now sew these together along the edges, we obtain a closed surface, which we call the *connected sum* of the surfaces M_1 and M_2 and denote by $M_1 \# M_2$. The process is shown in Fig. 2.21.

When we remove the inside of a triangle in a triangulation in each of the surfaces and sew them together along the edges, it is clear that $M_1 \# M_2$ "inherits" a triangulation from M_1 and M_2.

Apparently, the construction of a connected sum depends on how we cut the circular discs out of the surfaces in question. This, however, is only apparently the case, for it is possible, so to speak, to "pull" a circular disc around on a surface along an arbitrary curve; and if we pull the circular disc on M_1 around following a curve in M_1, then M_2 will follow along with it, and conversely, if we pull the circular disc on M_2 around following

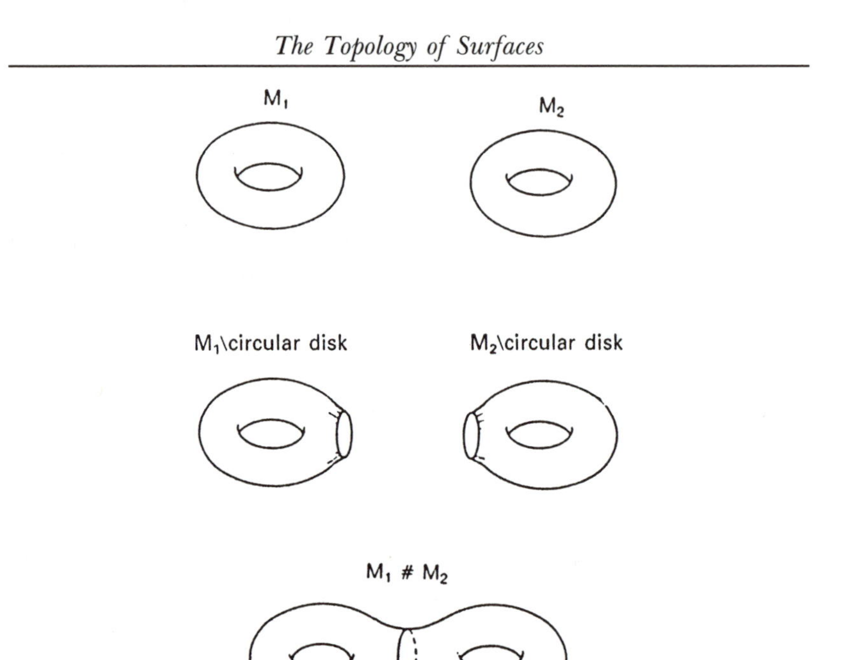

Figure 2.21.

a curve in M_2, then M_1 will follow along. If we imagine that the surfaces are made of rubber, we can thus "pull" M_1 and M_2 around in the connected sum until we obtain some other arbitrary positioning of the cut-out circular discs.

It is also easy to see that the sphere S^2 is a neutral element in the formation of a connected sum in the sense that for an arbitrary closed surface M, the connected sum $M \# S^2$ is equivalent to the surface M itself. The equivalence is demonstrated in Fig. 2.22, reminiscent perhaps of a weak spot in a bicycle inner tube.

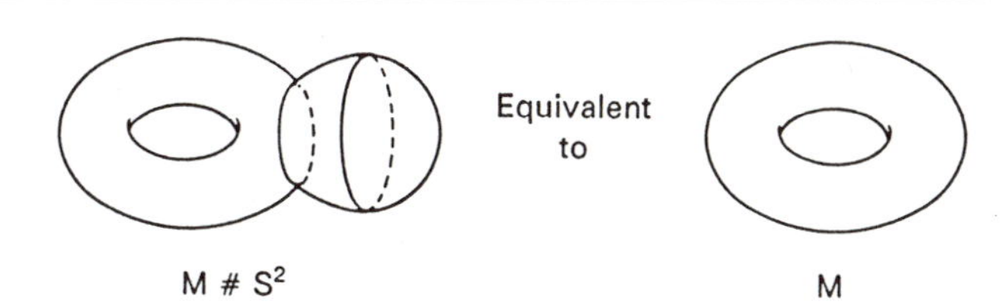

Figure 2.22.

By a *handle*, we mean the surface with a curved edge (a boundary) shown in Fig. 2.23, obtained by cutting a circular disc out of a torus T^2. If for a closed surface M we form the connected sum $M \# T^2$, we therefore also say we have *sewn a handle onto* M.

Now consider the projective plane $\mathbb{R}P^2$. By the definition of $\mathbb{R}P^2$, it follows that if we cut a circular disc out of $\mathbb{R}P^2$, we obtain a surface with an edge that is equivalent to a crosscap, which in turn—as stated in Proposition 2.1—is equivalent to a Möbius strip. We note this in the following:

Proposition 2.4. *For a closed surface M, the following operations produce equivalent closed surfaces*:

(1) *Formation of the connected sum* $M \# \mathbb{R}P^2$.

(2) *Cutting out a circular disc in M and sewing the edge diagonally.*

(3) *Cutting out a circular disc in M and sewing on a crosscap along the edge.*

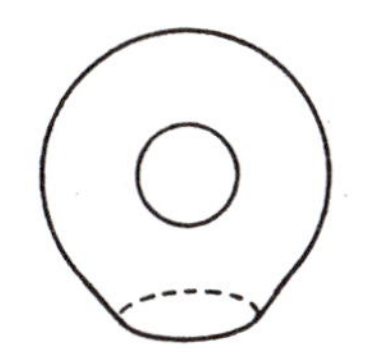

Figure 2.23. A handle.

(4) *Cutting out a circular disc in M and sewing on a Möbius strip along the edge.*

By comparing Propositions 2.3 and 2.4, we see that the Klein bottle K^2 is exactly the connected sum of two copies of $\mathbb{R}P^2$, in other words, that K^2 is equivalent to $\mathbb{R}P^2 \# \mathbb{R}P^2$.

We conclude this section with the following astonishing proposition:

(a)

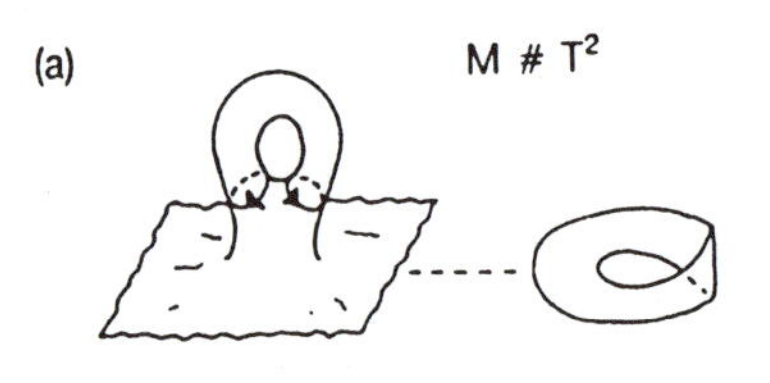

(d)

(b)

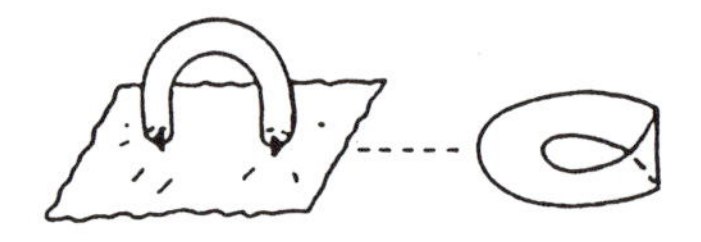

(e)

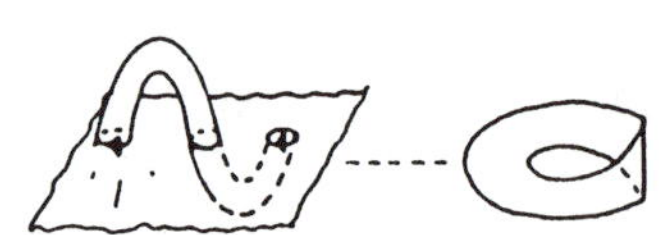

(c)

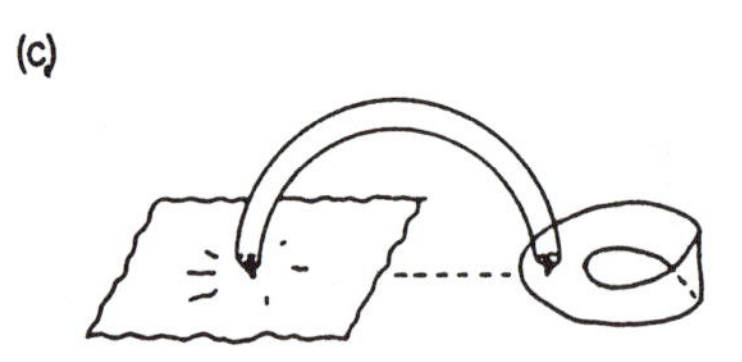

(f)

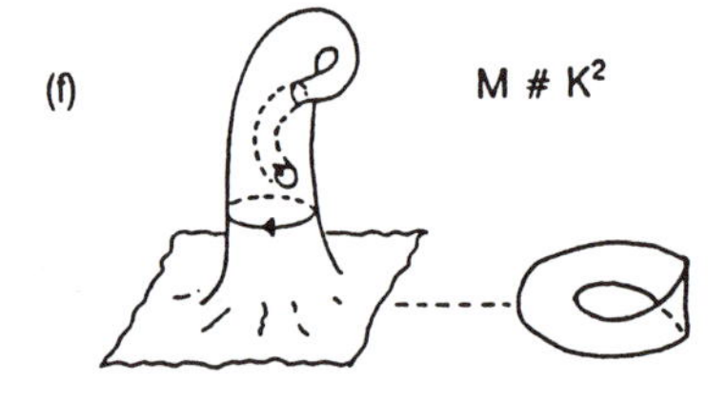

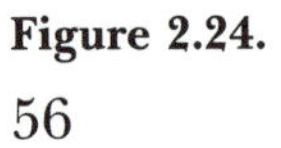

Figure 2.24.

Proposition 2.5. *Let M be an arbitrary non-orientable surface. Then the closed surfaces $M \# T^2$ and $M \# K^2$, formed by taking the connected sum of* M *with the torus* T^2 *and the Klein bottle* K^2, *respectively, are equivalent.*

We check this assertion by studying the series of pictures in Fig. 2.24. As usual, we imagine the surfaces to be made of rubber (of an unusually high quality). In the first picture (a), we have indicated $M \# T^2$ by showing the handle and the Möbius strip contained in the surface, since it is non-orientable. It is obvious that this picture corresponds to the picture in (b), which arises by "pulling" the handle "apart a little." By pulling the one end of the bended cylinder around on the surface, we obtain what is depicted in (c). If we now travel one circuit around the Möbius strip, we obtain what is depicted in (d). If we next pull back in the opposite direction from that of the transition from (b) to (c), we obtain what is depicted in (e). This picture corresponds to the picture in (f), which is, after all, precisely the connected sum $M \# K^2$. The assertion holds therefore, and it is a fact that $M \# T^2$ and $M \# K^2$ are equivalent.

6. Classification of Closed Surfaces

Generally, by *classification* of a collection of mathematical objects, one understands a division into classes comprising objects considered identical, or as mathematicians say, counted as *equivalent*. Moreover, a *model* for each equivalence class is to be given. A concept of equivalence shall, in other words, exist for those objects under consideration, and in order for a classification to be useful, the concept of equivalence must be such that it reflects essential characteristics of the objects and simultaneously gives rise to comparatively few equivalence classes. We have already encountered an example of a classification of a collection of mathematical objects, namely, the division of the regular polyhedra into five types. In this classification, we concerned ourselves with the kind of polygonal lateral surfaces, ignoring totally, for instance, the size of the polyhedra.

The collection of mathematical objects we shall now classify comprises the closed surfaces as described in Section 2.3. The underlying equivalence relation is *topological equivalence*, or *homeomorphism*, as mathematicians also call it. We say here that two closed surfaces M_1 and M_2 are (*topologically*) *equivalent* if a correspondence can be established between the points in M_1 and M_2 such that points lie closely together in one surface, if and only if the corresponding points lie closely together in the other surface. To add to this description, we must have a measure by which to judge the proximity in which points lie on a surface. Nor is this difficult to come by—the local charts on a surface, after all, describe local neighborhoods of points on the surface. In fact, we could define topological equivalence of surfaces by requiring not just that there be a correspondence between the points in the surfaces, but that this correspondence also can be extended to a correspondence between the systems of local charts on the surfaces.

It is clear that the sphere (a soccer ball) and the ellipsoid (a football) are equivalent. This can be seen by "pumping up" the ellipsoid till it is completely round, as shown in Fig. 2.25. The example in Fig. 2.26 shows that the equivalence relates only to the "inner" structure of the surfaces, and has nothing to do with their positioning in the surrounding space. A correspondence between the points on the knot and the torus in Fig. 2.26, which shows that the surfaces are equivalent, can be

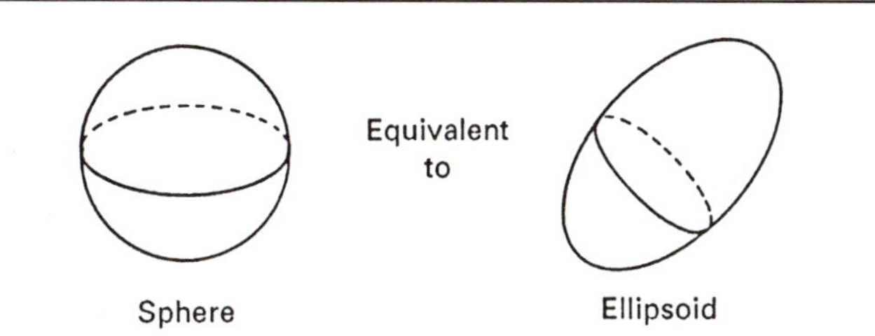

Figure 2.25.

constructed by running a meridian circle around each of the surfaces.

We now take up the description of the classification of the closed surfaces by first defining for every natural number $g = 1, 2, 3, \ldots$, two model surfaces F_g and U_g. The model surfaces are constructed as connected sums, respectively, of the torus T^2 and the projective plane $\mathbb{R}P^2$.

We begin by setting $F_1 = T^2$ and $U_1 = \mathbb{R}P^2$. For $g \geq 2$, F_g and U_g are defined consecutively from F_{g-1} and U_{g-1} by the formulas

$$F_g = F_{g-1} \# T^2,$$
$$U_g = U_{g-1} \# \mathbb{R}P^2.$$

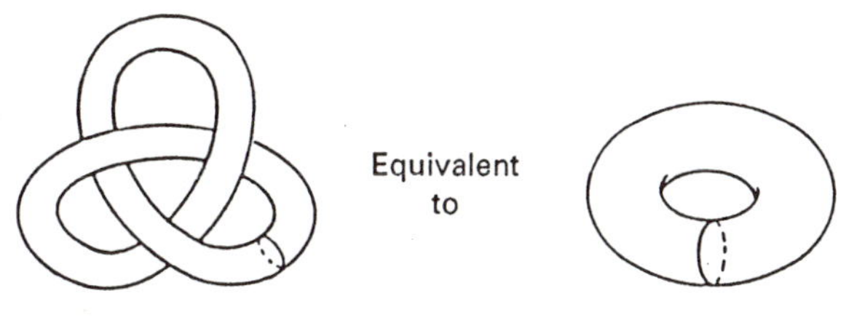

Figure 2.26.

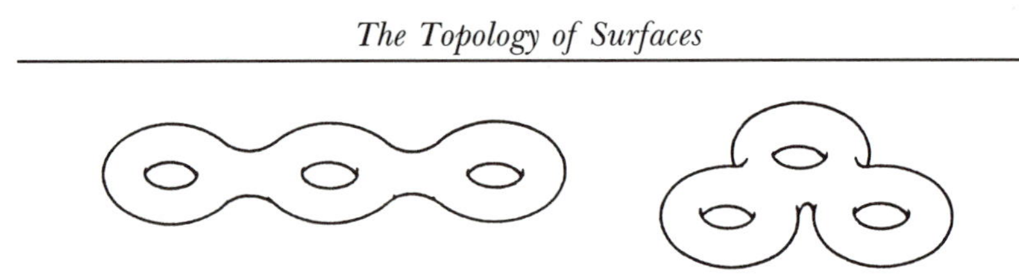

Figure 2.27. Equivalent versions of the model surface F_3.

From this, we derive

$$F_g = T^2 \# \cdots \# T^2 \ (g \text{ copies of } T^2),$$
$$U_g = \mathbb{RP}^2 \# \cdots \# \mathbb{RP}^2 \ (g \text{ copies of } \mathbb{RP}^2).$$

F_g is called a *model of the orientable closed surface of genus* $g \geq 1$. Correspondingly, U_g is called a *model of the non-orientable closed surface of genus* $g \geq 1$.

In Fig. 2.27, we show two equivalent versions of the model surface F_3. Imagining the surfaces as usual to be of extraordinarily high-quality rubber, we can always forcibly pull the handles around on the surfaces to any position.

The *genus* of a surface is a concept introduced by the great German mathematician Riemann (1826–1866) expressing how many times a cut can be made in a connected surface along a closed curve without self-intersections before the surface falls apart. Intuitively, it is clear—though not trivial to prove—that every cut in a sphere along a closed curve without self-intersections is going to separate the sphere into two pieces. Thus, one refers in this context to the sphere as *the closed surface of genus* 0. It is possible to make three such cuts on the surfaces in Fig. 2.27 without the surfaces' falling apart, but not four. Thus, these surfaces have genus 3.

By first observing that

$U_1 = \mathbb{R}P^2$,

$U_2 = \mathbb{R}P^2 \# \mathbb{R}P^2$, which is equivalent to the Klein bottle K^2,

it follows by employing Proposition 2.5 that

$$U_3 = \mathbb{R}P^2 \# \mathbb{R}P^2 \# \mathbb{R}P^2 \simeq \mathbb{R}P^2 \# K^2 \simeq \mathbb{R}P^2 \# T^2,$$
$$U_4 = \mathbb{R}P^2 \# \mathbb{R}P^2 \# \mathbb{R}P^2 \# \mathbb{R}P^2 \simeq K^2 \# K^2 \simeq K^2 \# T^2,$$

where the symbol $\simeq$ means the equivalence of surfaces.

In general, we derive that

$$U_{2n+1} \simeq \mathbb{R}P^2 \# T^2 \# \cdots \# T^2 \ (n \text{ copies of } T^2),$$
$$U_{2n+2} \simeq K^2 \# T^2 \# \cdots \# T^2 \ (n \text{ copies of } T^2).$$

It appears then that each of the surfaces F_g and U_g can be constructed by sewing handles on the sphere S^2, the projective plane $\mathbb{R}P^2$, or the Klein bottle K^2.

The classification theorem for closed surfaces says, in short, that two closed surfaces are topologically equivalent only when they have the same state of orientation and the same genus. Let us underscore that a closed surface is assumed to be connected and without boundary (edges). More precisely, we can formulate the classification theorem as follows:

The Classification Theorem for Closed Surfaces. *Every closed surface is topologically equivalent with just one of the surfaces* S^2, F_g, *or* U_g *for* $g \geq 1$, *and none of these surfaces are mutually equivalent.*

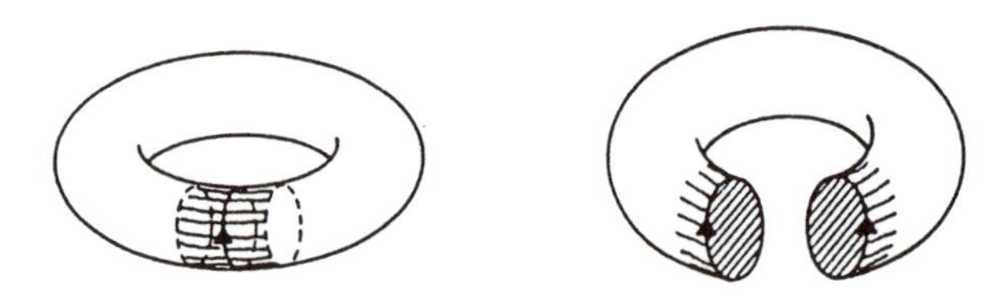

Figure 2.28. Surgery.

This theorem is a milestone in the history of topology, since here for the first time, as stated, a collection of non-trivial geometrical objects was successfully classified by its topological characteristics alone. The substance of the classification theorem for closed surfaces was known in one form or another by the end of the last century. Already in the 1860s, Möbius and Jordan had published proposals towards proofs of the classification of orientable surfaces, but their explanations suffer from the paucity of topological concepts developed by that time. Some maintain that a complete proof of the theorem was published in an article by the German mathematician Dehn and the Dane Heegaard in 1908. However that may be, one of the first complete proofs of the theorem was given in a book by the Hungarian mathematician Kerékjártó, published in 1923.

The main idea of the proof can be traced back to Riemann. Consider an arbitrary closed surface M. The idea is now to transfrom M stepwise by cutting along closed curves without self-intersections and sealing the resulting holes. The cutting and the sealing proceed until a further cut will divide the surface. The closed curve of the cut that we employ in each step of the process will be contained in a strip that is equivalent either to a cylinder, as in Fig. 2.28, or to a Möbius strip. At each step in the process, we cut out the entire strip, and we obtain thereby—corresponding to the two possibilities—either two holes, or one hole, with "circles" as edge curves. If we then seal the two holes (or respectively, the one hole) with circular discs, we will have reached the next step in the process. The case with two holes is shown in Fig. 2.28. Notice that we can at once return to the immediately preceeding step in the process by sewing on, respectively, a handle (forming a connected sum with a torus), or a crosscap (forming a connected sum with a projective plane). There are two points we shall not prove here. The first is that after a finite number (the genus of the surface) of steps, we reach a surface where a further cut will divide the surface. The second is that such a surface is a sphere. Now we reverse the whole process, and work our way back by, at each step, forming a connected sum with a torus or a projective plane, and as indicated before, we return by this means to the original surface M. In rough outline, this proves the theorem. Modern topology dubs the cutting and sealing processes we have undertaken *surgery*.

7. Higher-Dimensional Manifolds and Poincaré's Conjecture

After their success with the classification of closed surfaces in the beginning of this century, topologists turned their attention to higher-dimensional manifolds. Since discussion of these developments is more mathematically demanding, this section may optionally be skipped for now.

A *closed n-dimensional topological manifold* M is a connected geometrical object, without a boundary, that can be locally described with n coordinates. For example, a three-dimensional closed manifold is a connected geometrical object locally resembling the normal three-dimensional space we live in. Locally, we can think of a piece of a three-dimensional manifold as a room (a box) where we determine the position of a point by three coordinates, e.g., the ordered set of three real numbers, made up of the distances from the front wall, from one side wall, and from the floor. In the general case, the local model of an n-dimensional manifold is the n-dimensional real number space $\mathbb{R}^n$, whose points are ordered sets of n real numbers. As in the case of surfaces, the n-dimensional manifolds may have additional structure as *smooth* or *piecewise linear* manifolds. Corresponding to the two-dimensional sphere S^2 in three-dimensional space, there exists a standard n-dimensional sphere S^n in the $(n + 1)$-dimensional number space $\mathbb{R}^{n+1}$ for every dimension n. The one-dimensional sphere S^1 is of course the circle in the plane, whose points are at a distance 1 from a fixed origin in the plane.

In dimension 2, it is entirely crucial for the classification of closed surfaces that we have a simple characterization of the sphere S^2, namely as a closed surface in which every cut along a closed curve on the surface (without self-intersections) divides it into two pieces. This last characteristic may also be formulated by a requirement that every closed curve on the surface can be contracted to a point on the surface; we say in this case that the surface is *simply connected*. The equivalence class of the sphere S^2 is therefore characterized by the fact that a closed surface is topologically equivalent to S^2 if and only if it is simply connected.

In dimension 3, the problem of classification runs up against a serious obstacle in that there exists no theorem that characterizes S^3 in the same way in which we were able to characterize S^2. In 1899, the great French mathematician Henri Poincaré (1854–1912) formulated a conjecture that every simply connected, closed three-dimensional manifold as described earlier is topologically equivalent (homeomorphic) with S^3. Despite many attempts to prove this conjecture, it remains an unsolved problem of mathematics. A positive solution to Poincaré's conjecture will lead to a complete classification of closed three-dimensional manifolds. An especially significant work on topology in three dimensions, honored with the Fields Medal in 1982, was carried out by the American mathematician William Thurston, who has established a whole program to classify three-dimensional manifolds. In the case of a three-dimensional manifold, one can always assume, after a possible shifting of charts, that the charts overlap smoothly or piecewise linearly. Thus, from the standpoint of classification, as in dimensions 1 and 2, the topological, the piecewise linear, and the smooth manifolds coincide in dimension 3 as well.

In dimension 4, there was great success in the 1980s. There is a generalized form of Poincaré's conjecture in every dimension n. In 1983 (announced in the summer of 1981), the American mathematician Michael Freedman succeeded in establishing a thorough classification of simply connected, closed four-dimensional topological manifolds, including in particular the solution to the generalized Poincaré conjecture in dimension 4. This achievement brought him the Fields Medal at the International Congress of Mathematicians held at the University of California at Berkeley in 1986.

There is a strong parallel in many ways to the classification of closed surfaces, so we shall return briefly to the surfaces. By an embedding of the circle into a surface M, we understand a closed curve without self-intersections on M; in short, we speak of a circle on M. Two circles on M are said to belong to the same deformation class if they can be deformed into

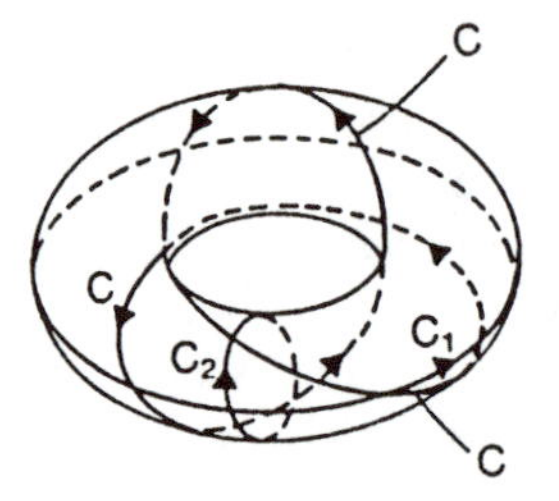

Figure 2.29. Two essentially different non-contractible circles C_1 and C_2 on a torus that constitute a basis for the circles on the torus, together with a circle C that has the intersection number 3 with C_1 and intersection number 2 with C_2.

each other through circles; if they cannot be deformed into each other through circles, they are considered essentially different. It can be proved that every surface M has a basis for its circles, which means a set of essentially different circles $C_1, \ldots, C_r$ on M that cannot be contracted to points, so that the deformation class of any other circle C on M is completely determined by the number of points of intersection with each of the circles $C_1, \ldots, C_r$ (where every point of intersection is reckoned as the case may be as $+1$ or -1 depending on the direction of rotation of the circles). The number of circles in a basis of circles on M does not depend on the set and is called the *rank* of M. On a sphere, all circles may be contracted to points, so the sphere has rank 0. The torus has a basis for its circles with two essentially different embeddings of the circle that cannot be contracted to points, for example, the circles C_1 and C_2 shown in Fig. 2.29, and the torus thus has rank 2. On a closed, orientable surface of genus g there is in complete generality a basis for the circles with $2g$ such embeddings of the circle, and the rank of an orientable closed surface of genus

g is thus exactly $2g$. As we have seen in Section 2.6, a closed orientable surface is classified by its genus g. Since the genus determines the rank, the closed orientable surfaces are thus also classified by their rank.

Now to the four-dimensional manifolds. With every simply connected, closed four-dimensional manifold M, there is similarly associated a positive integer called the *rank* of M. The rank of M is in this case the number of essentially different two-dimensional spheres that cannot be contracted to a point in a basis of the two-dimensional spheres in M. In order to classify the four-dimensional manifolds, we need in addition to the rank a quadratic schema of integers in which the number of rows and columns in the schema equals the rank of the manifold. In such a schema, we list how many times—again, counted with sign determined by orientations—any two arbitrarily chosen two-dimensional spheres in a basis for the two-dimensional spheres embedded in M intersect each other. For every rank, there are of course an infinite number of different schemas, but they are not all associated with a manifold. It is comparatively easy to determine which schemas come from four-dimensional manifolds, but it is difficult to determine whether two different schemas come from the same four-dimensional manifold.

There are an astonishing number of different topological types of simply connected, closed four-dimensional manifolds. Thus, of rank 40, there are more than 10^{51} different topological types—an unimaginably large number greater than the square of the estimated number of particles in the universe. A unique contribution was made by the young English mathematician Simon Donaldson in 1983, when by utilizing methods from gauge field theory, to which we shall return in Chapter 5, he proved that there are many fewer topological types of smooth, simply connected, closed four-dimensional manifolds. Combining Donaldson's and Freedman's results, one obtains, stated briefly, that every smooth, simply connected, closed four-dimensional manifold (with particular requirements concerning the quadratic schema) regarded topologically can be written as a connected sum of copies of the four-dimensional sphere S^4, the manifold $S^2 \times S^2$, the complex projective plane $\mathbb{C}P^2$ (which corresponds to the real projective plane $\mathbb{R}P^2$, when we work over the complex numbers instead of over the real numbers), plus one further complicated but well-known four-dimensional manifold. This achievement naturally enough led to Donaldson's receiving the Fields Medal at the same time as Freedman in 1986.

By combining his own results with Donaldson's, Freedman, together with the celebrated American topologist R. C. Kirby, made a striking discovery that startled the mathematical world. To begin with, Freedman and Kirby discovered that the four-dimensional real number space $\mathbb{R}^4$ permits more than one smooth structure, i.e., that there exists more than one atlas of local charts of $\mathbb{R}^4$ that overlap smoothly, and that cannot by shifting charts (maintaining smooth overlapping) be transferred to the same atlas of local charts of $\mathbb{R}^4$. In all other dimensions, the smooth structure of the n-dimensional real number space $\mathbb{R}^n$ is uniquely determined. It has emerged meanwhile that there actually exist an infinite number of smooth structures on $\mathbb{R}^4$.

The problem of classification in dimensions ≥ 5 was solved in the 1960s. The greatest obstacle here, too, was the generalized Poincaré conjecture in these dimensions. This conjecture for smooth manifolds was proved by the American mathematician Steven Smale in 1960, an achievement that brought him the Fields Medal at the International Congress of Mathematicians held in Moscow in 1966. With a little nudge, the generalized Poincaré conjecture for piecewise linear manifolds in dimensions ≥ 5 also follows from Smale's results. This was also proved independently with methods from piecewise linear topology by Stallings and Zeeman 1961–62. The generalized Poincaré conjecture was proved for topological manifolds in dimensions ≥ 5 by the English mathematician Newman in 1966.

As announced by Markov at the International Congress of Mathematicians in Edinburgh in 1958, the completely general problem of dividing the closed manifolds in a given dimension n into equivalence classes of topologically equivalent manifolds proves to belong to the class of unsolvable problems. (It is curious, and at first glance somewhat surprising, that there are problems in mathematics that can mathematically be proven unsolvable.) Already in 1953, however, the French mathematician René Thom—in a pioneering work that was honored with the Fields Medal in 1958—classified closed, smooth manifolds by using a much less refined equivalence concept called *cobordism*.* A similar classification after the cobordism type of closed topological manifolds was carried out in the early 1970s in a collaboration between the American mathematicians Brumfield and Milgram and the Danish mathematician Ib Madsen.

* Two n-dimensional manifolds are called *cobordant*, if together they constitute the boundary of a manifold in dimension $n + 1$. For example, all closed, orientable surfaces are cobordant with the sphere, since the model surface F_g and the sphere S^2 together constitute the boundary of the three-dimensional manifold obtained by taking a pretzel with g holes (the solid version of the model surface F_g) and cutting a solid ball out of the interior of the pretzel.

We have discussed smooth manifolds, piecewise linear manifolds, and topological manifolds in the preceding. It turns out that every smooth manifold can be given a piecewise linear structure. Correct proofs of this were given by Cairns in 1934 and J. H. C. Whitehead in 1940. In a groundbreaking work that in a certain sense initiated the study of topology of smooth manifolds (differential topology), John Milnor proved in 1956 that on the seven-dimensional sphere S^7, there exist several different smooth structures that fit together with the topological structure. In actual fact, there exist precisely 28. For this achievement, Milnor was honored with the Fields Medal at the International Congress of Mathematicians in Stockholm in 1962. Then in 1960, Michael Kervaire proved the existence of a 10-dimensional piecewise linear manifold whose piecewise linear structure does not derive from a smooth structure. Finally, in 1969, after groundbreaking work of Dennis Sullivan, Kirby and Siebenmann completely clarified the connection between topological and piecewise linear manifolds. In this connection, they demonstrated among other things that topological manifolds exist that do not permit a piecewise linear structure.

In all the investigations of manifolds cited previously, increasingly refined versions of the surgical techniques we encountered in Section 2.6 have played a decisive role. Among the pioneers in the development of these techniques must be noted the American mathematicians John Milnor, William Browder, and Dennis Sullivan, the English mathematician C. T. C. Wall, and the Russian mathematician S. Novikov. Presently, one of the leading experts in the field is the Danish mathematician Ib Madsen.

3

THE TOPOLOGY OF CATASTROPHES

IT IS SAFE TO SAY that in *catastrophe theory*, the French topologist René Thom hit on a catchy name for the theory whose foundations he laid toward the end of the 1960s. Along with Thom, catastrophe theory has another leading figure in the English topologist Christopher Zeeman, and it is probably the latter who has done the most toward popularizing the theory, and toward its being, for a time at least, widely discussed. Thom and Zeeman do not fully agree as to its scope and potential. While Thom conceives catastrophe theory as more or less a construct of ideas—a way of thinking—capable of describing only qualitative aspects of phenomena, Zeeman, in the hope of proving the theory experimentally in the long run, has attempted to introduce quantitative elements as well.

The object of catastrophe theory is to construct topological models of phenomena that exhibit sudden jumps (catastrophes). Models have been proposed for phenomena as varied as fetal development, aggression, declarations of war, phase transitions, the stability of ships, and the collapse of a stock market. The theory should be regarded as supplemental to the deterministic models of systems that evolve in space and time, described mathematically by ordinary differential equations.

The underlying mathematics builds on the theory of singularities of differentiable mappings, which in all its essentials was developed after 1960. The founder of the theory was my mathematical grandfather, the American mathematician Hassler Whitney (1907–1989).* Other leading figures include, in particular, Thom, the American mathematician Mather, and the Russian mathematician Arnold.

Towards the end of the 1970s, a battle was raging between ardent supporters of catastrophe theory and their very outspoken critics. The seas seem now to have calmed, and the theory is entering a phase where it can certainly contribute toward the understanding of our world, though it no longer gives the impression that it is an all-encompassing theory; and this may be to the good, for a theory that explains everything explains nothing.

In this chapter, we offer an introduction to catastrophe theory. Though the presentation is relatively straight-forward, it will, especially in Sections 7 and 8, require some background in mathematics. The mathematics in these sections is included to show clearly that the underlying mathematics has never been in question, only some of its applications, of which we also present a few of the more controversial. For the more philosophical aspects of the theory, and for further applications, the reader is referred to the two books by the theory's leading figures, R. Thom and E. C. Zeeman, cited in the bibliography. At several points in the presentation, the reader is referred directly to these books.

* The author's Ph.D. advisor at the University of Warwick, England was Professor James Eells, whose own Ph.D. advisor in turn was Hassler Whitney.

1. The Origin of Catastrophe Theory

Catastrophe theory was created by the French mathematician René Thom in the 1960s. Underlying it are phenomena within the purview of biology, physics, and other sciences, where small shifts in the determinants can cause sudden changes (jumps, or catastrophes).

Embryology encounters many such sudden changes. The initial developmental phases of the fetus, for instance, run as follows. Development starts with the *fertilized ovum*. Then begins the process of *mitosis*, in which the fertilized ovum divides into two cells, which each in turn divide into two, and so on, until it has divided into up to 1,000 smaller cells, occupying *in toto* the same volume as the single original ovum. In this process, a hollow called the *blastula cavity* forms, followed by a process called *gastrulation* in which the divided ovum enters the blastula cavity, assuming the shape of a tiny pot. In space-time (four dimensions), these changes can be seen as a series of sudden jumps.

Another example is phase transition, the everyday experience encountered when a substance suddenly shifts from one state to another. A small change in the temperature around the freezing point of water can cause water to freeze into ice, or ice to melt into water.

Inspired by phenomena such as those just mentioned, Thom asked himself the bold question whether there are mathematical limits on how such jumps (catastrophes) can occur. In the course of exploring this question among others, Thom found there are just seven so-called elementary catastrophes in space-time. These do not explain everything; rather, they are just the beginning of a more general theory.

Thom was born in 1923, and as mentioned earlier, in 1958 he won the Fields Medal for his pioneering work in topology. He is Professor at the Institut des Hautes Études Scientifiques, a research institute at Bures-sur-Yvette just south of Paris. Thom has described his ideas about catastrophe theory in the very speculative but fascinating book, *Stabilité Structurelle et Morphogénèse*, cited in the bibliography. He has focused especially on applications in biology, notably in *embryology* and *morphogenetics* (the study of the origin and development of a biological form or structure throughout its existence).

Due in part to its provocative name, catastrophe theory has captured public attention, and has been discussed in publications such as *Newsweek*, *New Scientist*, *Scientific American*, and also in *The New York Times* and other dailies in several countries. The theory has been controversial, and still is to some extent, and though criticism is subsiding, mathematicians Hector Sussman and Raphael Zahler have attacked it in print, and a critical report on catastrophe theory by journalist Gina Koleta in the magazine *Science*, Vol. 15 (April, 1977), bears the title "Catastrophe Theory: The Emperor Has No Clothes."

But no one disputes the fact that the theory has stimulated the development of mathematics significantly. In particular, catastrophe theory has occasioned a whole series of new discoveries regarding differentiable functions of several variables; we cite among others an important theorem of the French mathematician Malgrange: "The Preparation Theorem".

2. Singularities: Mappings of the Plane into the Plane

The mathematical background for Thom's development of catastrophe theory is the theory of singularities of differentiable mappings. To illustrate some of the ideas in the theory

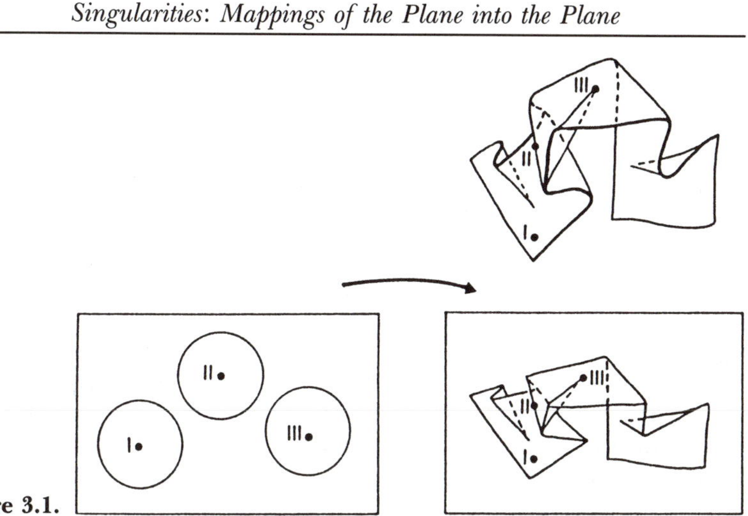

Figure 3.1.

of singularites, we shall now describe a theorem by Whitney from 1955 dealing with singularities of mappings of the plane into the plane. Before Thom, Hassler Whitney pioneered the development of a theory of singularities.

We can imagine a mapping of the plane into the plane as resulting from laying a piece of (crepe) paper over a piece of paper of the same shape. Crumpling, shrinking, and stretching are all permitted. Singularities arise where the picture is crumpled up (Fig. 3.1).

A little experimentation should convince you that if we watch a small neighborhood in the mapping, perhaps following a preparatory perturbation as small as we wish, only three local types will occur: *regular point*, *fold point*, *and cusp point* (Fig. 3.2).

These are the three local types that are generally unavoidable when we map the plane into the plane. These local types

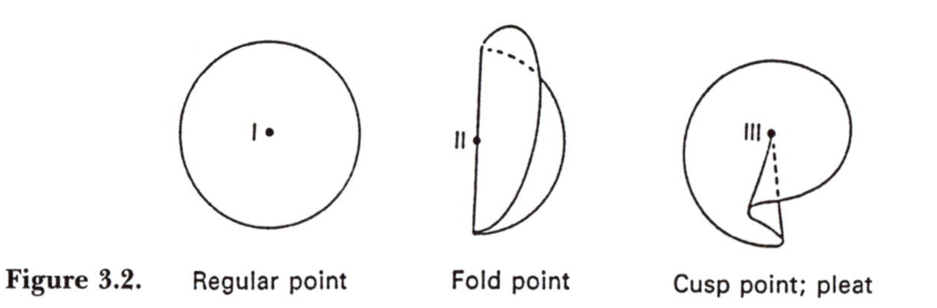

Figure 3.2. Regular point Fold point Cusp point; pleat

are also stable, i.e., they do not disappear with small perturbations. Whitney's theorem states that we can approximate every (infinitely often differentiable) mapping of the plane into the plane as closely as we wish with a mapping displaying only these three local types.

The fold and the cusp appear among the so-called elementary catastrophes. To clarify the ideas underlying Thom's catastrophe theory, let us first consider the simplest case, namely, the fold catastrophe.

3. The Fold Catastrophe

Behind everything in Thom's theories lies the concept of *stability*. Stability (mathematically, structural stability) is a fundamental condition that Thom thinks should be placed on all models of phenomena taken from reality. His fundamental viewpoint is that natural phenomena possess a high degree of stability, for if they did not, small disturbances would end their existence, and thus they could not be observed.

We shall now offer an example of how an unstable situation is embedded in a stable situation (mathematically, a universal unfolding of a singularity).

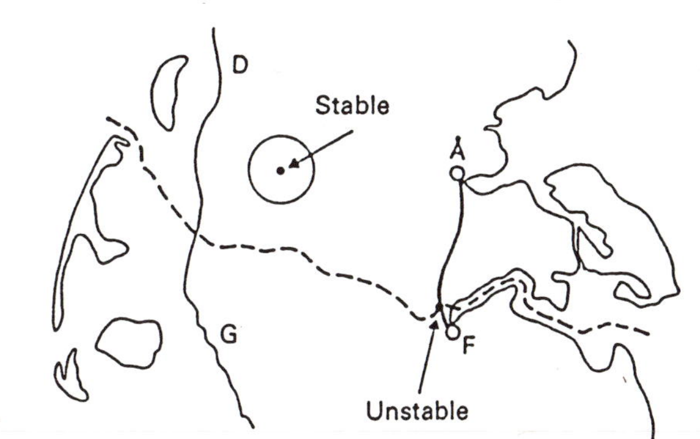

Figure 3.3.

Example 3.1. Let us consider two countries bordering each other, for example, Denmark and Germany. See Fig. 3.3.

If a point is chosen at random on the map, it is most likely that it will fall in one of the two countries. If a point falls within one of the two countries, a whole neighborhood of the point will belong to the country in question. It is therefore a *stable* situation to land within one country in the sense that "one resembles one's neighbors." If the point falls on the border, the situation is unstable—some of the points in a neighborhood belong to the one country, and others to the other country. The border points are *catastrophe points* (drastic changes occur in crossing over).

Let us choose two towns, one in each country, Åbenrå and Flensburg, say, and look at the connecting roads between the two towns. On each of these roads we will have to cross the border. (A catastrophe will occur.) If we choose any particular road between the two towns, we see that on nearby roads the trip from one town to the other will occur in the same way as along the chosen road. Thus, the catastrophe occurs consistently along all roads. We have gotten the unstable situation "unfolded" into a stable situation.

A border crossing illustrates the *fold catastrophe*. Birth is another example—and death. For the individual human being, it is a catastrophe (an unstable situation) to die, but since all humans die, the catastrophe "dying" occurs consistently (stably) throughout humanity. An example of a fold catastrophe from daily life: a teacup knocked to the floor.

The catastrophe most often cited is the cusp catastrophe, which we shall now describe.

4. The Cusp Catastrophe

By way of prelude to a discussion of the cusp catastrophe, it will be useful to construct and play with a simple machine devised by Christopher Zeeman. We follow the instructions on pages 8–12 of Zeeman's book (listed in the bibliography).

To make the machine, we need two rubber bands, two thumbtacks, half a matchstick, a piece of stiff cardboard, and a small wooden board. (You might imagine yourself in a Little Rascals movie.) Cut a disc out of the stiff cardboard with a diameter approximately the length of one of the unstretched rubber bands. Affix both rubber bands to a point H near the edge of the disc. A simple method is to make a small hole in the disc at H, pushing loops of the rubber bands through the hole, and fastening them by sticking the half matchstick through the loops and pulling tight, as in Fig. 3.4a. Now fasten the center of the disc to the wooden board with the thumbtack

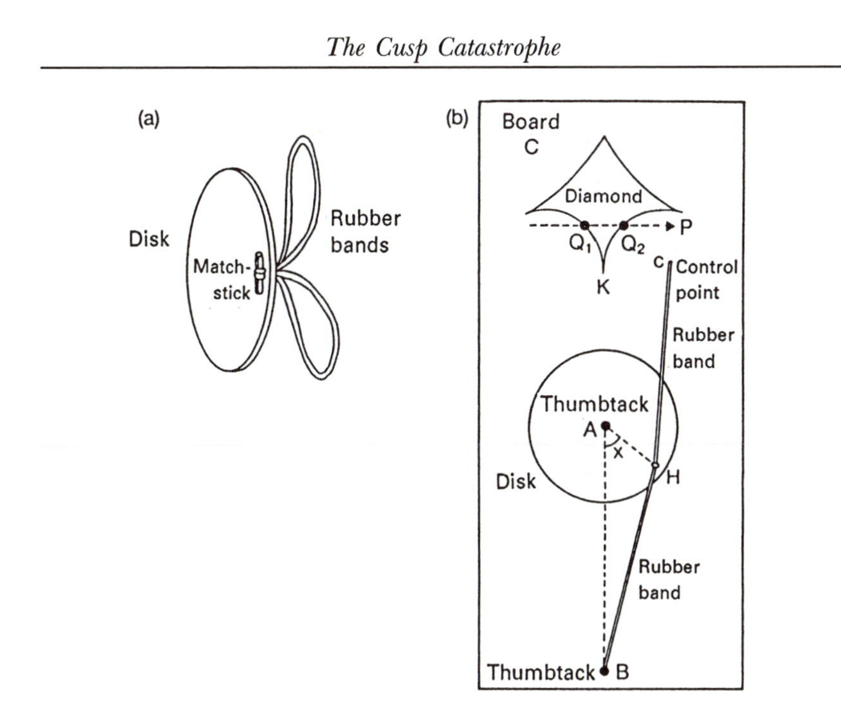

Figure 3.4. Zeeman's catastrophe machine.

A (so that the rubber bands are above and the matchstick underneath), making sure the disc can rotate freely. Press the other thumbtack B into the wooden board so that the distance AB is approximately twice the length of the unstretched rubber band, and slip one of the rubber bands over B. The machine is now as shown in Fig. 3.4b and ready for use.

Hold the free end of the other rubber band against the wooden board; wherever this is held is *control point* c. *Control plane* C is thus the whole wooden board. The *state* of the machine is given by the position of the disc, which we can measure by the angle of rotation x. As control point c is moved evenly, state x follows, except that now and then instead of moving evenly, it makes sudden jumps. Each time it jumps, we mark the corresponding control point with a ball-point pen. After a short time, we will have enough points marked to connect them with a concave diamond-shaped curve K with four cusps, as shown in Fig. 3.4b.

Playing with the catastrophe machine leads us to the following discoveries: Some areas in the control plane admit two possible states of the system (called *bimodality*); sudden jumps can occur; two paths running close together in the control plane can lead to widely different results (called *divergence*); and finally, the system displays *hysteresis*, by which we mean that the return jump after a sudden jump does not occur at the same control point Q_1 where the jump took place, but somewhere out along the opposite branch of the diamond curve at Q_2. (See Fig. 3.4b.)

The dynamics of the catastrophe machine—what brings the wheel to a position of equilibrium—is steered by the potential energy of elasticity in the rubber bands. The energy of elasticity is determined by Hooke's law. We have two control variables and one state variable (the angle of rotation). Now mark the angle(s) of rotation associated with a given control point perpendicular to the control plane at a height corresponding to the size(s) of the angle(s) of rotation. We obtain thereby a surface in space representing equilibrium positions that in the neighborhood of a cusp looks like Fig. 3.5.

The intermediate surface over the area in the control plane inside the cusp represents an inaccessible location for the wheel. It corresponds to states where the energy has a local maximum.

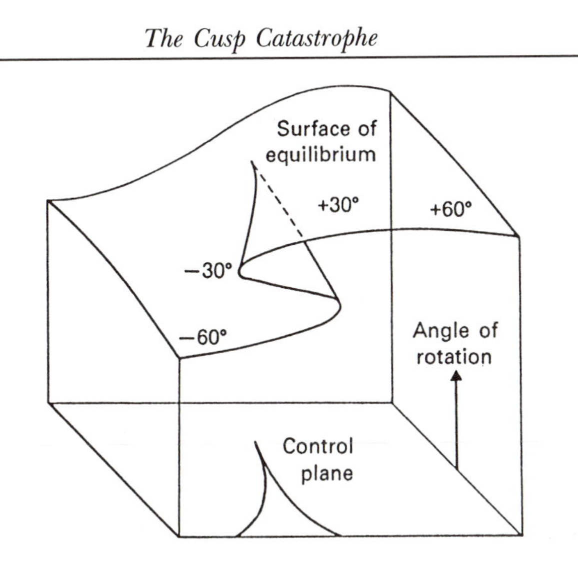

Figure 3.5.

For systems with two control variables and one state variable, governed by a potential function, Thom's theorem proves that when we mark the state variable perpendicular to the control plane, the resulting surfaces will in "almost all" cases locally be smooth (regular points), resemble a fold, or resemble a cusp (as in Fig. 3.5).

The cusp represents a so-called *cusp catastrophe*. In the vicinity of the cusp, the system to which the surface belongs displays *bimodality*, *sudden jumps*, *divergence*, and *hysteresis*.

In a system in which we can identify two control variables and one state variable, and where one or more of the characteristics of bimodality, sudden jumps, divergence, and hysteresis occurs, there is reason to turn to the cusp catastrophe and try to identify the other characteristics as well.

We shall now attempt briefly to specify the mathematics of the preceding.

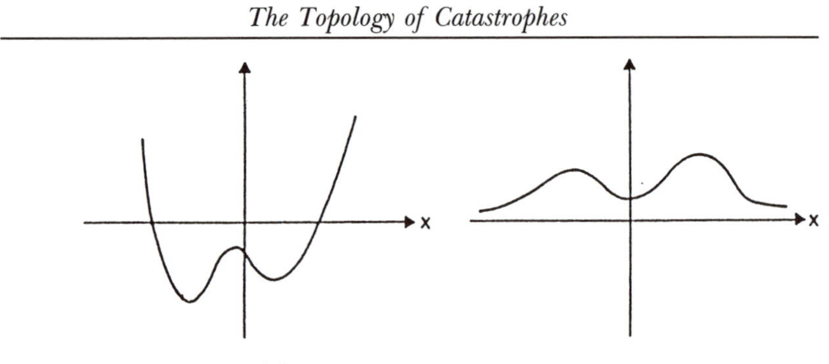

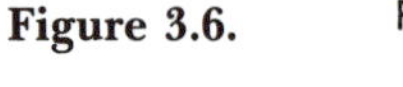
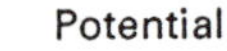

Figure 3.6. Potential Probability distribution

5. Thom's Theorem for Systems with Two Control Variables and One State Variable

Consider a system that can be described by a single state variable $x \in \mathbb{R}$ and can be controlled (affected) via two control variables $(u, v) \in \mathbb{R}^2$.

For fixed values of the control variables $(u, v) \in \mathbb{R}^2$, we let $f_{(u,v)}: \mathbb{R} \rightarrow \mathbb{R}$ be a potential function, or a probability distribution, that governs the dynamics of the system, i.e., leads the system to equilibrium when the system is affected by the control $(u, v) \in \mathbb{R}^2$. The meaning of $f_{(u,v)}(x)$ thus is as follows:

(i) In potential formulation, $f_{(u,v)}(x)$ is the potential of the system in the state $x \in \mathbb{R}$ under the control $(u, v) \in \mathbb{R}^2$.

(ii) In probability formulation, $f_{(u,v)}(x)$ is the probability that the system will be in state $x \in \mathbb{R}$ under the control $(u, v) \in \mathbb{R}^2$.

In Fig. 3.6, we have shown the graph for $f_{(u,v)}$.

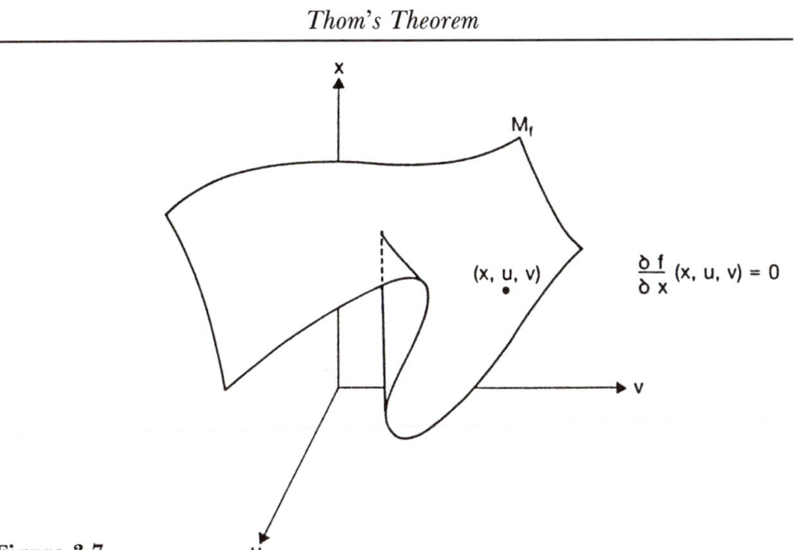

Figure 3.7.

The functions $f_{(u,v)}: \mathbb{R} \to \mathbb{R}$ are put together into a function

$$f: \mathbb{R}^3 = \mathbb{R} \times \mathbb{R}^2 \to \mathbb{R}, \quad f(x, u, v) = f_{(u,v)}(x).$$

We assume now that f is infinitely often differentiable and form

$$M_f = \left\{ (x, u, v) \in \mathbb{R}^3 \middle| \frac{\partial f}{\partial x}(x, u, v) = 0 \right\}.$$

M_f is a surface in $\mathbb{R}^3$ that contains the possible states of the system (Fig. 3.7). In potential formulation, we consider minima, while maxima are excluded. In probability formulation, we consider maxima, while minima are excluded.

Theorem 3.1. (*Thom's theorem for two control variables*). *For "almost all" functions* $f(x, u, v): \mathbb{R}^3 \to \mathbb{R}$, *we obtain by the preceding construction a surface* M_f *that contains only regular points, fold points, or cusps.*

6. Some Applications of the Cusp Catastrophe

Let us now consider some applications of the cusp catastrophe based on the work of E. C. Zeeman. The applications are controversial; nevertheless, they describe the ideas underlying catastrophe theory rather well. In particular, the model for aggression (Example 3.2) has been the object of criticism, but the reader should decide for himself. Probably a certain intellectual shift is required to accept mathematical models of this kind.

Example 3.2 (Model of *aggression*). In connection with what follows, the reader should definitely also consult Zeeman's own presentation, on pages 3–8 of his book.

Let us consider *aggression* in a dog with respect to a human being. The state of the dog is measured on a one-dimensional scale of aggression running from flight through neutrality to attack. To obtain a "continuous" scale, aggression might perhaps be measured by the concentration of adrenalin in the blood.

According to the late Austrian animal behaviorist (ethologist) Konrad Lorenz, the dog's aggression is governed by its *fear* and its *rage*. According to Lorenz, these two parameters are in mutual conflict. Fear alone leads to flight. Rage alone leads to attack. When fear and rage are both present, both flight and attack are possible outcomes. Neutral behavior is

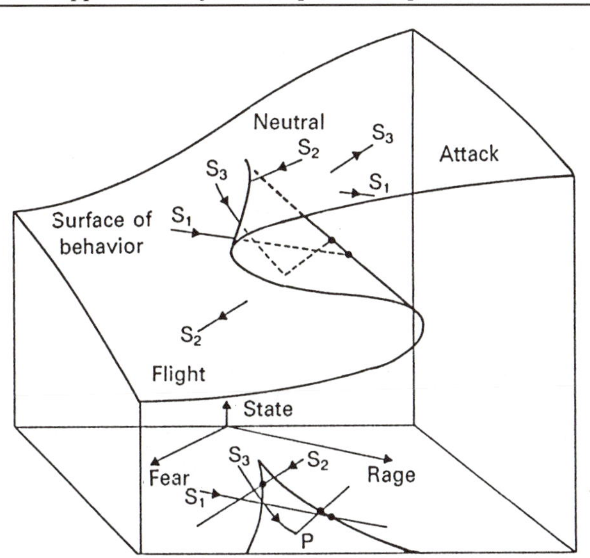

Figure 3.8.

the least probable outcome. Again, according to Lorenz, it is possible to read the two parameters, fear and rage, in the dog's face. In response to fear, the ears are laid back; to rage, the teeth are bared.

Since we have two control variables, and since there are areas with two possible outcomes (bimodality), Thom's theorem suggests that we experiment with the cusp catastrophe as a model.

Since the parameters fear and rage are in mutual conflict, the cusp is to be placed between the coordinate axes as in Fig. 3.8.

The path S_1 describes a typical progression: Mounting rage in the dog leads finally to attack. The model suggests that with the cooling of the rage along path S_1, the shift to flight occurs only at a lower level of rage than that occasioning the original attack. This hysteresis effect can also be observed in practice.

The path S_2 is equally typical: Mounting fear in the dog leads at last to flight.

Path S_3 is interesting. At the crisis point P, the dog has reached quite a high level of rage, but it is nonetheless still in retreat. Then the human being relaxes (turns his back, perhaps), and the dog's fear diminishes, but since it remains fixed at the high level of rage, we run off over the edge and the dog suddenly leaps to attack when least expected.

We can, as mentioned, also observe bimodality. In a situation where the dog is both strongly enraged and very fearful, neutral behavior is the most improbable—the dog will either attack or flee. In the model, neutral behaviour corresponds to the intermediate surface, which is inaccessible.

It is interesting to note that the catastrophe model of aggression links together neurophysiology (processes in the dog governing aggression such as adrenalin secretion) and psychology (our observations of the dog's reactions). Neurophysiology takes place in the state space; psychology in the control space.

The phenomena in the preceding model can also be observed in a heated discussion between two people. The control parameters here are fear and rage in the dominated party, and the variable of state is his aggression. If the dominated person feels cornered, his reaction will be unpredictable, and can swing from violent aggression to a sudden outburst of tears. We recognize this volatility in the phenomenon known as hysterical weeping.

Example 3.3 (Illustration of the proverb "*Haste makes waste*").

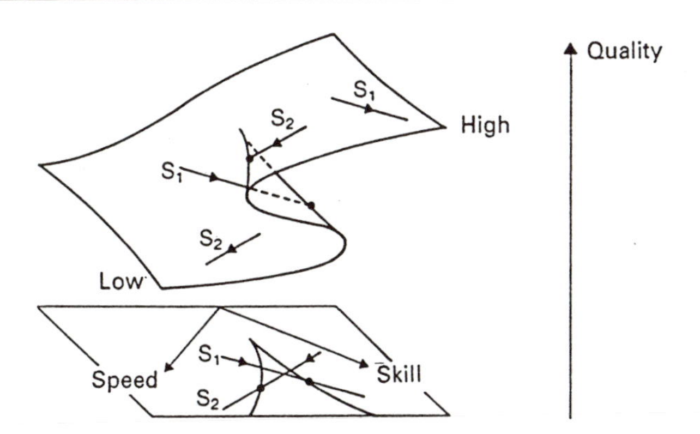

Figure 3.9.

Assume we wish to practice a certain skill, say, typing. We measure *quality* by, e.g., the number of errors per 100 strokes.

The quality depends on the typist's *skill* and the typist's *speed*.

It is clear that great skill and low speed lead to high quality, and that little skill and great speed lead to low quality. In the event of both great skill and great speed, mediocre (neutral) quality is the least probable outcome—either the typist outdoes himself and we get good quality, or the typist runs amok and we get poor quality.

Since we have a situation with two control variables (skill and speed) and one state variable (quality), and since there are areas in the control plane with bimodality, Thom's theorem inspires us to experiment with the cusp catastrophe as a model (Fig. 3.9).

Along path S_1, we observe how practice of the skill "suddenly" leads to high quality—one has "got the hang of it." Along path S_2, we observe how mounting pressure for increased typing speed leads "suddenly" to low quality.

The hysteresis effect is observable here, too. Once pushed over the edge along path S_2 and lapsed into poor quality, one has to lower one's speed significantly to get going again.

The jump in the training process is perhaps clearer still if we think of a child learning to ride a bicycle. It is experienced as truly a leap when the child is suddenly able to hold himself upright on the bicycle.

Example 3.4 (Model of *economic growth*). *The rate of economic growth* is indicated on a one-dimensional scale. GNP, rate of unemployment, etc. are important factors in determining a country's rate of economic growth.

Influencing the rate of economic growth, a country has two important control parameters, namely, deflation (tightening credit, compulsory savings, etc.) and devaluation (lowering one's own currency's value).

Deflation slows the wheels of the economy and thus results in a low rate of growth. Devaluation makes it cheaper for foreign countries to buy the goods one produces and can thus lead to economic growth. It is well-known that "sudden" jumps in the rate of growth can occur. Therefore, we shall experiment once more with the cusp catastrophe as a model. Since the two control parameters are in mutual conflict, the cusp is to be placed between the coordinate axes, as in Fig. 3.10.

Assume now that the country's government wishes to go from state T_1 to state T_2 over a certain period.

We see that path I leads to a high rate of growth (as occurred in France in 1968), while path II leads to a low rate of growth (as occurred in England in 1967). A possible explanation lies in the fact that deflation slows the economy so that nothing is produced. If devaluation follows, demand will most likely increase, but the warehouses are empty, so there is nothing to sell.

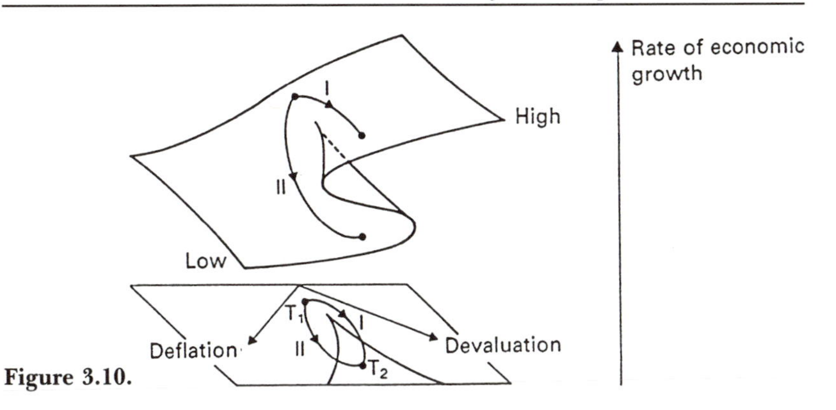

Figure 3.10.

The model serves as a good warning. Even if observable catastrophes occur neither along path I nor along path II, they lead to widely different results. The reason is that one has been on a catastrophe surface that cannot be observed directly.

The model also warns against uncritically employing statistical models, for statistics serve to smooth out data. In the process, though, it is easy to smooth out a cusp catastrophe, for instance, and thereby overlook an important phenomenon.

7. The Mathematics behind the Models of Catastrophe Theory

We shall now try to describe the mathematical models that general catastrophe theory occupies itself with, and simultaneously look at the simplifying hypotheses that lead to elementary catastrophe theory. The section does require some familiarity with differential calculus, and may be skipped over.

In constructing mathematical models of systems that evolve in time, systems of differential equations are often employed.

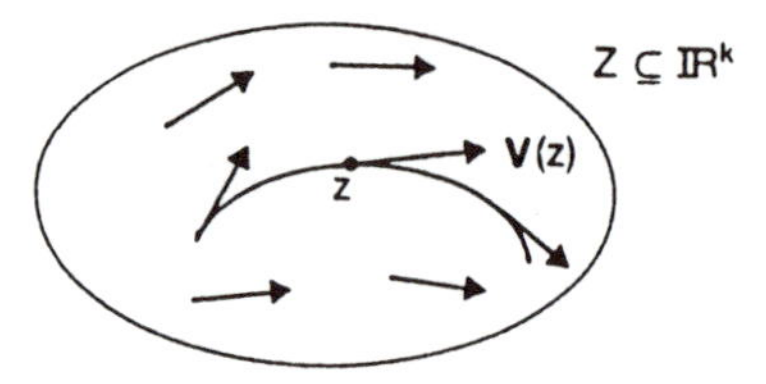

Figure 3.11.

Consider a system that can be described by k real parameters $z = (z_1, \ldots, z_k) \in \mathbb{R}^k$. For example, the system might be a cell where $z_1, \ldots, z_k$ are the concentrations of the different substances contained in the cell. Let us say that the possible states of the system are described by the points z in the open subset $Z \subseteq \mathbb{R}^k$. Assume that the system's evolution (the dynamics of the system) is described by the system of ordinary differential equations

$$\frac{dz_i}{dt} = f_i(z_1, \ldots, z_k), \qquad i = 1, \ldots, k, \tag{3.1a}$$

or equivalently as in Fig. 3.11, by

$$\text{a vector field } \mathbf{V} \text{ on } Z \subseteq \mathbb{R}^k. \tag{3.1b}$$

The connection between the system of differential equations and the vector field is that the vector at the point $z \in Z$ is given by $\mathbf{V}(z) = (f_1(z), \ldots, f_k(z))$.

The system's evolution in time is obtained by following the vector field $\mathbf{V}$. At a point $z \in Z$, the arrow's direction tells the direction in which the system is evolving, and the arrow's length indicates the speed with which it is evolving.

It is almost clear that the evolution is determined when we know the state of the system at a given point in time t_0, that is, we know the starting point $z_0 \in Z$, corresponding to t_0. From a mathematical standpoint, this is contained in a deep-seated theorem from mathematical analysis (the existence and uniqueness theorem of systems of ordinary differential equations). The model is thus *deterministic*.

Model building along these lines has had many successes in physics. Among others, we can cite Newton's laws and Maxwell's laws, to which we shall return in Chapters 4 and 5.

Two difficulties arise, however, with models of this sort. First, a system of differential equations is not exactly created to describe situations where small variations can cause sudden jumps. Small changes of data in a deterministic model yield small changes in the end results. More seriously, one is often dealing—especially in biology and in the social sciences—with systems containing a great number of variables, which on the one hand are not readily observable, and on the other cannot easily be subsumed into a smaller number of "macroscopic" variables.

Let us assume, for instance, that we have a system such as in Eqs. (3.1), with k very large, where there are a smaller number of "important" variables (which we perhaps can control), $y_1, \ldots, y_r$, which are functions of the variables $z_1, \ldots, z_k$,

$$y_i = g_i(z_1, \ldots, z_k), \qquad i = 1, \ldots, r. \tag{3.2}$$

If now the variables $y_1, \ldots, y_r$ are the only ones that are significant to us, it would of course be nice if we could get a system of differential equations in the y's or a vector field in y space consistent with Eqs. (3.1), which governed the evolution of the variables y. This is, however, not normally possible.

The situation is not artificial. Consider for example the motion of a particle in space. Here we might be interested only in the particle's spatial coordinates, but we are forced to involve the inertial coordinates (the coordinates of velocity) as new variables to describe the system's evolution. This procedure has been known in physics since Newton's time. There are indications that analogous tricks will not work in, say, biology.

The principal objective of catastrophe theory is to study a system of differential equations

$$\frac{dz_i}{dt} = f_i(z_1, \ldots, z_k), \qquad i = 1, \ldots, k, \tag{3.3a}$$

or equivalently,

$$\text{a vector field } \mathbf{V} \text{ on } Z \subseteq \mathbb{R}^k, \tag{3.3b}$$

together with functions

$$y_i = g_i(z_1, \ldots, z_k), \qquad i = 1, \ldots, r, \tag{3.4}$$

with the object of deriving as much information as possible about what occurs in the space described by the coordinates $y_1, \ldots, y_r$.

The system of differential equations, or the vector field, in Eqs. (3.3) is here regarded as being of lesser interest. Only the phenomena observable in the y coordinates are considered of interest.

A program like this is much too general to admit hope of success. As a first simplification, let us assume, therefore, that the functions $g_1, \ldots, g_r$ are simply the coordinate functions for the restriction to $Z \subseteq \mathbb{R}^k$ of the projection

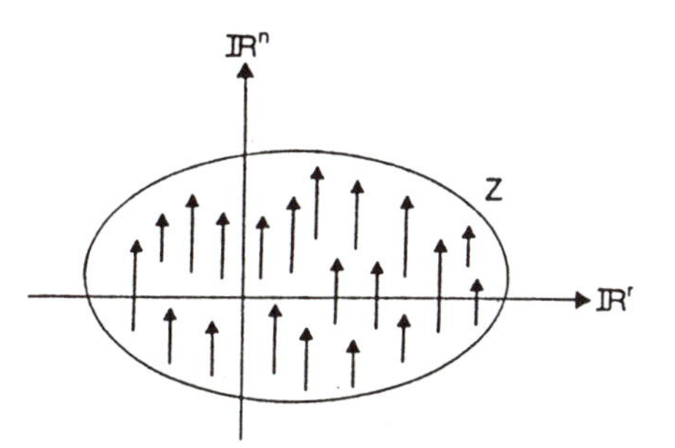

Figure 3.12. The vector field **V** in Eq. (3.3b) is directed along R^n.

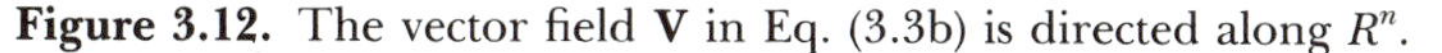

$$\mathbb{R}^k = \mathbb{R}^n \times \mathbb{R}^r \rightarrow \mathbb{R}^r, \qquad z = (x, y) \rightarrow y,$$

where $z = (x, y) = (x_1, \ldots, x_n, y_1, \ldots, y_r) \in \mathbb{R}^n \times \mathbb{R}^r = \mathbb{R}^k$.

Further simplification is achieved by assuming that the dynamics (the vector field) in Eq. (3.3b) is mainly directed along R^n, that is, changes in accordance with $x = (x_1, \ldots, x_n)$ occur rapidly relative to changes in accordance with $y = (y_1, \ldots, y_r)$. See Fig. 3.12.

With these simplifications, we are now led to consider models of the form

$$\pi: X \times Y \rightarrow Y, \qquad (x, y) \rightarrow y,$$

where $X \subseteq \mathbb{R}^n$ and $Y \subseteq \mathbb{R}^r$ are sets of local coordinates in R^n and R^r, respectively (or more generally, smooth manifolds X and Y of dimensions n and r, respectively), together with a vector field **V** on $X \times Y$ along X.

The study of such models constitutes what we understand by *general catastrophe theory*.

For the moment, the mathematical foundations are too weak to attack this general formulation of the problem. In his book, therefore, Thom largely restricts himself to considering models where the vector field **V** is a gradient field. The study of such models is called *elementary catastrophe theory*.

We assume now, therefore, that there exists a smooth (differentiable of class C^∞) function

$$\mathrm{f}\colon X \times Y \to \mathbb{R},$$

such that the vector field **V** on $X \times Y$ along X is given by

$$\mathbf{V}(x, y) = \left(\frac{\partial \mathrm{f}}{\partial x_1}(x, y), \ldots, \frac{\partial \mathrm{f}}{\partial x_n}(x, y), 0, \ldots, 0\right).$$

For fixed $y \in Y$, we can consider the function $\mathrm{f}_y\colon X \to \mathbb{R}$, where $\mathrm{f}_y(x) = \mathrm{f}(x, y)$, as a potential function (or a probability distribution) for a system described by the points in X under the external condition $y \in Y$. The points in X are called *state variables* (or internal variables), and the points in Y are called *control variables* (or external variables). The system will thus stay in minimum points (maximum points in the case of the probability model) for f_y.

We therefore consider the set of stationary points

$$\mathrm{M_f} = \left\{(x, y) \in X \times Y \,\middle|\, \frac{\partial \mathrm{f}}{\partial x_1}(x, y) = \cdots = \frac{\partial \mathrm{f}}{\partial x_n}(x, y) = 0\right\}.$$

The equations in the definition of $\mathrm{M_f}$ place n restrictions on the points in the $(n + r)$-dimensional point set $X \times Y$. We can

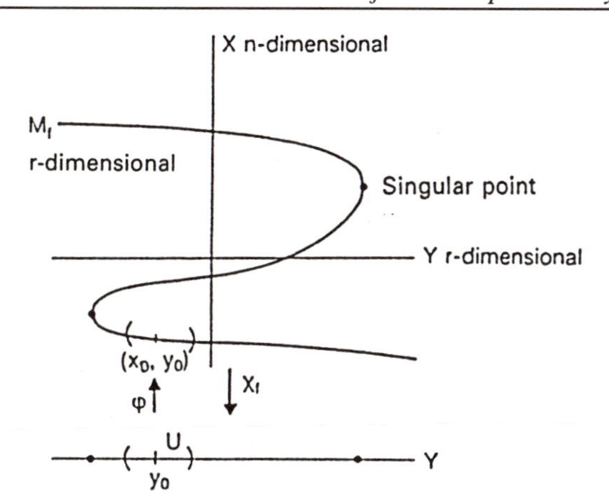

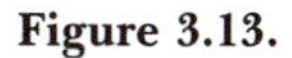
Figure 3.13.

therefore expect that M_f will become r-dimensional (the dimension of Y). See Fig. 3.13.

It turns out also that M_f for "almost all" smooth functions f: $X \times Y \to \mathbb{R}$ becomes a nice smooth r-dimensional point set in $X \times Y$ (precisely, an r-dimensional submanifold). For such f, we consider the mapping

$$\chi_f\colon M_f \to Y, \qquad (x, y) \to y,$$

defined by restriction of the projection π: $X \times Y \to Y$ to M_f.

Provided that the determinant

$$\det\left[\frac{\partial^2 f}{\partial x_i \partial x_j}(x_0, y_0)\right] \neq 0 \text{ in a point } (x_0, y_0) \in M_f,$$

a deep-seated theorem from mathematical analysis (the implicit function theorem) shows that there exists an open neighborhood U of $y_0 \in Y$ and a smooth function φ: $U \to X$, such that $y \to (\varphi(y), y)$ yields a parameterisation of M_f in a neighborhood of $(x_0, y_0) \in M_f$. This neighborhood of $(x_0, y_0) \in M_f$ is mapped smoothly onto the neighborhood U of $y_0 \in Y$ by the mapping χ_f. When the control y is varied within U, the system will therefore continuously adjust itself smoothly and without jumps.

We are therefore interested in the singular points for χ_f, that is, points on M_f, where a neighborhood is crumpled up by the map χ_f. The singular points for χ_f are exactly the points $(x_0, y_0) \in M_f$, where the determinant vanishes:

$$\det\left[\frac{\partial^2 f}{\partial x_i \partial x_j}(x_0, y_0)\right] = 0.$$

When the control y in this situation is varied in a neighborhood of $y_0 \in Y$, the system can for certain control values choose among several possible states, and can thus display jumps. The singular points for χ_f define the so-called *elementary catastrophes*.

Prompted by the preceding, we introduce the notation

$$\Delta_f = \left\{(x, y) \in M_f \,\middle|\, \det\left[\frac{\partial^2 f}{\partial x_i \partial x_j}(x, y)\right] = 0\right\}$$

and

$$D_f = \chi_f(\Delta_f).$$

The mapping χ_f is called the *catastrophe mapping*, and D_f is called the *catastrophe set.* Possible jumps in the state of the system occur along D_f.

If we look only at χ_f in a neighborhood of a singular point for $\chi_f: M_f \to Y$, it turns out that when the dimension of Y is ≤ 5, there is only a finite number of different topological (geometric) models of χ_f. If the dimension of Y is ≤ 4, there are only seven topologically different models. These seven models are the celebrated seven elementary *catastrophes in space-time* discovered by Thom.

For dimension $Y \leq 5$, there are 11 different models. For dimension $Y \geq 6$, there are infinitely many different models in every dimension.

Every elementary catastrophe that occurs for a certain dimension of Y occurs again in all higher dimensions by merely adding "trivial" variables in Y. The smallest dimension of Y for which an elementary catastrophe first occurs is called its *codimension*. Thus, we obtain the following table:

Table 3.1

Codimension	1	2	3	4	5	6
Number of elementary catastrophe	1	1	3	2	4	∞

It is worth noting that the dimension of X plays quite a subordinate role.

Mathematicians like Thom, Mather, and Arnold have made especially important contributions to the classification of the elementary catastrophes.

When we consider only a neighborhood of a fixed singular point for χ_f belonging to the function $f: X \times Y \to \mathbb{R}$, we can restrict ourselves to considering a smooth function

$$f: \mathbb{R}^n \times \mathbb{R}^r \to \mathbb{R},$$

defined in an open neighborhood of $(0, 0) \in \mathbb{R}^n \times \mathbb{R}^r$, with $f(0, 0) = 0$ and with $(0, 0) \in \mathbb{R}^n \times \mathbb{R}^r$ as a singular point for the associated catastrophe mapping $\chi_f: M_f \subset \mathbb{R}^n \times \mathbb{R}^r \to \mathbb{R}^r$. We note that $(0, 0) \in M_f$ when

$$\frac{\partial f}{\partial x_i}(0, 0) = 0, \qquad i = 1, \ldots, n,$$

and that $(0, 0) \in M_f$ is a singular point for χ_f when the determinant

$$\det\left[\frac{\partial^2 f}{\partial x_i \partial x_j}(0, 0)\right] = 0.$$

The function $f_0(x) = f(x, 0) = \mathbb{R}^n \to \mathbb{R}$ is called a *singularity*, and the function $f: \mathbb{R}^n \times \mathbb{R}^r \to \mathbb{R}$ is called an *unfolding* of the singularity f_0. Thom thinks of the singularity f_0 as a potential function that contains information to be revealed or decoded. If r is exactly the codimension for the elementary catastrophe belonging to f, the function f is called a *universal unfolding* of the singularity f_0. In a sense that can be made precise, a universal unfolding of f_0 contains all unfoldings of f_0.

8. The Seven Elementary Catastrophes in Space-Time

In this section, we present in Figs. 3.14–3.20 the seven elementary catastrophes in codimension ≤ 4. For each of the seven elementary catastrophes, we indicate a smooth function $f: \mathbb{R}^n \times \mathbb{R}^r \to \mathbb{R}$, with $(0, 0) \in \mathbb{R}^n \times \mathbb{R}^r$ as the singular point for χ_f, which describes the catastrophe presented. The dimension n is the lowest possible, and r is the codimension for the elementary catastrophe. Deviating from the notation of Section 7, in Figs. 3.14–3.20, x and y both denote state variables (the coordinates in $\mathbb{R}^n$), while control variables (the coordinates in $\mathbb{R}^r$) are denoted by t, u, v, and w.

The names of the catastrophes are translations of Thom's original French names. We also indicate some of Thom's interpretations of the catastrophes, in space and time.

Elementary catastrophes

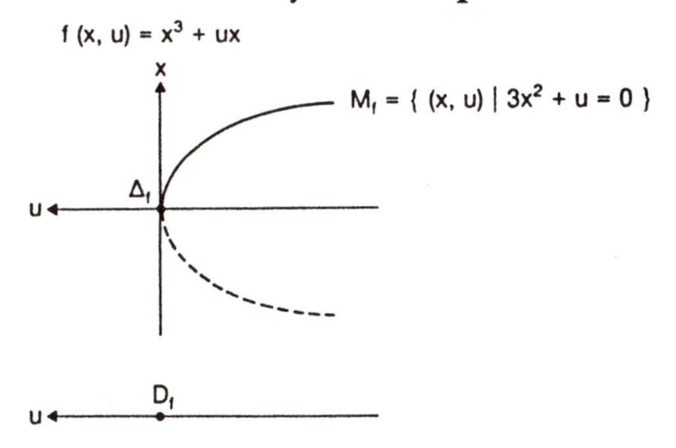

Figure 3.14. Codimension 1: The fold. Space: The boundary, the end. Time: Starting, finishing.

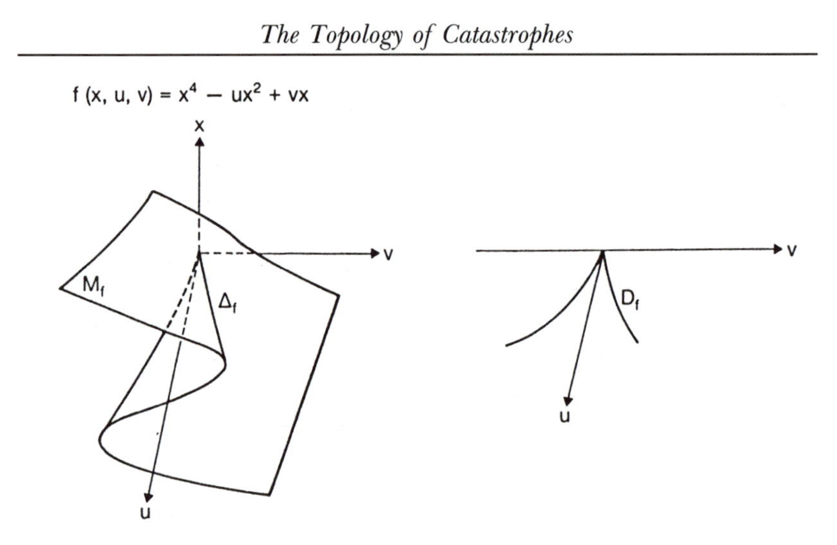

Figure 3.15. Codimension 2: The cusp (the Riemann-Hugoniot catastrophe). Space: A pleat, a (geological) displacement. Time: Separating, unifying, capturing, generating, altering.

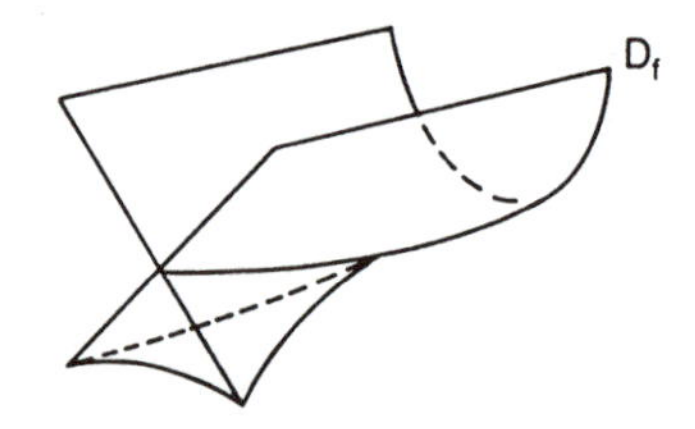

Figure 3.16. Codimension 3: The swallow-tail. Space: A fissure, a furrow. Time: Splitting, sawing, tearing apart.

$$f(x, y, u, v, w) = x^3 + y^3 + wxy - ux - vy$$

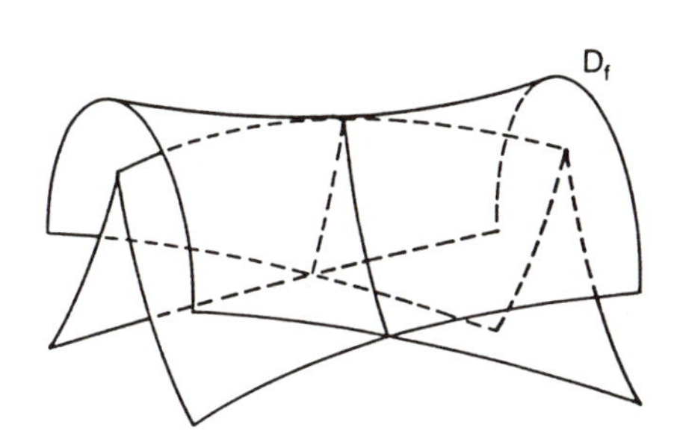

Figure 3.17. Codimension 3: The hyperbolic umbilic. Space: Wave crest, vault (arc). Time: Breaking (of a wave), collapsing, swallowing up.

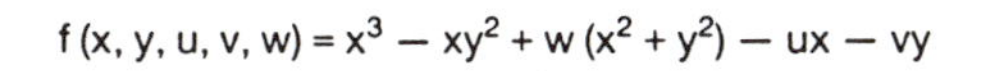

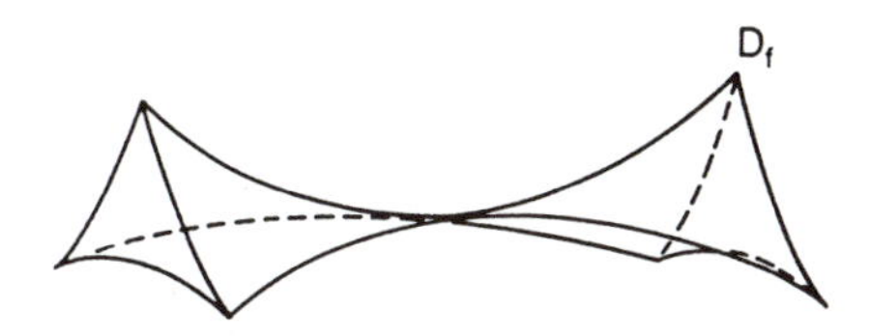

Figure 3.18. Codimension 3: The elliptical umbilic. Space: Needle, apex, hair. Time: Drilling or filling (a hole), stabbing.

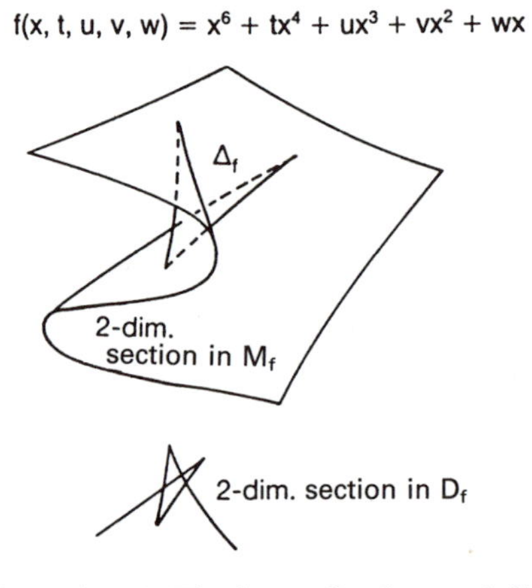

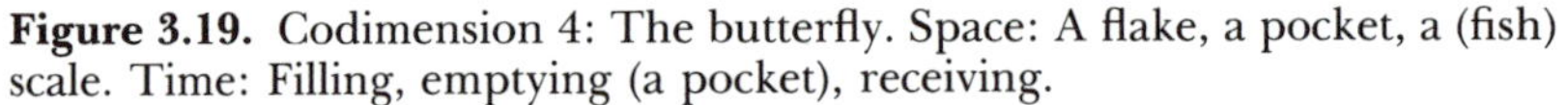

Figure 3.19. Codimension 4: The butterfly. Space: A flake, a pocket, a (fish) scale. Time: Filling, emptying (a pocket), receiving.

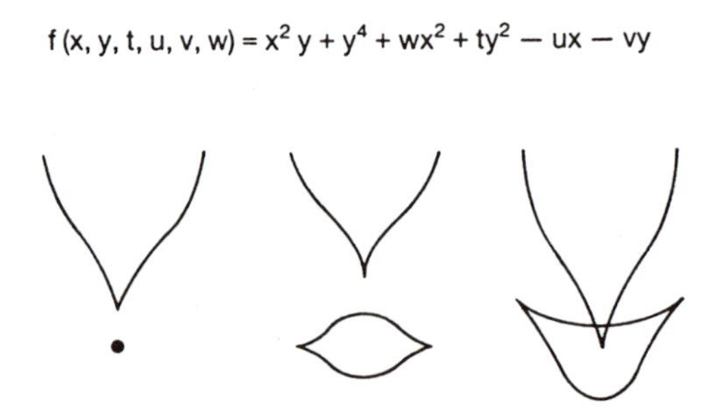

Figure 3.20. Codimension 4: The parabolic umbilic. These are some two-dimensional sections in D_f. Space: Jet (of a wave), mushroom, mouth. Time: Breaking (for a jet), opening or closing one's mouth, piercing, cutting, pinching, taking, expelling, throwing.

For a more thorough description and for more diagrams, the reader is referred to Chapter 5, pp. 55–100 in Thom's book, cited in the bibliography.

9. Some General Remarks concerning Applications

Thom places great significance on the three umbilics. In anatomy, *umbilic* is the term for *navel*, and according to Thom, the three umbilics describe, among other things, those catastrophes by which a newly created organism separates itself from its progenitor. (See Thom's book, p. 81.)

An umbilic has two interior parameters (state variables). Thom believes that the three umbilics will play a leading role in phenomena of refraction in hydrodynamics (refraction of waves, refraction of jet-streams). In biology, he believes, they govern the morphogenesis of absorption phenomena such as phagocytosis. An example of phagocytosis is familiar from the process by which white blood cells swallow bacteria. Another example of phagocytosis is found among certain one-celled organisms that excrete through the pinching off of vacuoles (blisters filled with cytoplasm) by the cell membranes from which wastes are diffused by osmosis. Thom thinks that the three umbilics have significance for neurology as well.

By now, articles have been published suggesting applications of catastrophe theory to numerous fields. Without attempting to be comprehensive, we present some of the more acknowledged applications.

Physics

Here, we must first cite a series of works on optics by Thom, Berry, Jänisch, Guckenheimer, and Arnold.

Engineering

(i) Two English professors of engineering named Thompson and Hunt (initially independent of Thom's theory) discovered that the hyperbolic and the elliptical umbilics are of great importance in the *theory of elasticity*. There are also several papers by Zeeman on this subject.
(ii) A paper on the *stability of ships* (Zeeman) also looks intriguing.

Biology

(i) The heart attack (Zeeman).
(ii) Culmination of slime-molds (Zeeman).
(iii) Darwin's law of natural selection (Dodson).
(iv) A whole series of still unexplored ideas put forward by Thom.

Economics

Here, we must first cite a controversial paper about the *collapse of a stock market* by Zeeman.

Medicine

In a model of the illness *anorexia nervosa*, Zeeman employs the butterfly catastrophe.

A comprehensive bibliography, which includes references to the articles mentioned here, can be found in the recommendable book by T. Poston and I. Stewart cited in the bibliography. With this reference, we close our discussion of catastrophe theory.

GEOMETRY AND THE PHYSICAL WORLD

Ever since classical antiquity, geometry has played a central role in man's understanding and description of the immediate physical world, and in his construction of models of the universe. Conversely, ideas from the physical world are of fundamental significance in the formation of mathematical intuition. Thus, the Egyptians' need for land surveying was a crucial precondition to their development of the basic concepts of geometry in the plane. With the many contributions of the Greeks, classical geometry reached its first peak around 300 B.C. when Euclid set down what then was known, in his famous *Elements*.

In the earliest models of the world, the Earth was flat; beyond the horizon yawned an abyss. The Pythagoreans, however, declared open-mindedly that the Earth was a sphere. Even if this was the dominant conception from then on among astronomers, the myth of a flat Earth persisted to haunt popular conception, right up to the time when Columbus's voyage across the Atlantic in 1492 rendered this viewpoint untenable. In the wake of Vasco da Gama's circumnavigation in 1519, the conclusion was inescapable that the world is without edge and runs back into itself—in short, that it is round. In the impressive photographs of the Earth from space, we have gotten a powerful impression of its shape and a clear message as to its limits.

The Earth was conceived as the center of the universe until the middle of the 16th century when Copernicus concluded on aesthetic grounds that the sun must be the center. At the time of Copernicus, a mechanical model had been constructed permitting description with great precision of the motions of the planets known at that time, but the system had grown so complicated that Copernicus discarded the geocentric model in favor of the more aesthetically pleasing heliocentric model, which had already been proposed hypothetically by Aristarchus of Samos (c. 310–230 B.C.), active in the Greek astronomical school in Alexandria. On the basis of Tycho Brahe's observations, Kepler published early in the 16th century his three laws of planetary motion, the first of which is formulated in purely geometrical terms, namely, that the planets move in elliptical orbits with the sun at one of the focal points. Towards the end of the 17th century, Newton, in his famous *Philosophiae Naturalis Principia Mathematica*, formulated the laws of nature that now constitute the foundations of classical mechanics. It was a great triumph for Newton that from these laws he could derive Kepler's laws.

When one reflects on how mankind in the course of millenia evolved a model of the world that described the known universe very precisely, but which as we now know was completely erroneous, one may well remain somewhat skeptical of computer models not based on real understanding, or whose ultimate goal is not the quest for explanation.

1. On Mathematics and Its Greek Legacy

Around 600 B.C., the Greeks seriously began to consider mathematics as a logical structure and as a tool for learning more about nature. It is difficult to say why; all one can say for sure is that from about this time, the Greeks were convinced that in all essentials nature is rationally organized, and that all the phenomena of nature pursue a precise and stable plan; indeed, a mathematical plan. The human brain has an extraordinarily large capacity, which if exploited in the study of nature, can reveal and utilize this rational mathematical plan.

Thus, the Greeks became the first people to attempt to reason their way to explanations of natural phenomena. The mathematical arguments of the Greeks were largely geometric, and while they explored things, they developed aids from geometry, such as Euclidean geometry, so that others might more readily reach the frontiers and help to achieve new conquests. In this way, the Greeks contributed to the founding of the methodology of modern science.

According to legend, Thales (c. 640–545 B.C.), who lived in the Greek town of Miletus in Asia Minor, proved several theorems of Euclidean geometry, but we have no reliable testimony about this. It is certain, however, that he and his contemporaries speculated on the organization of nature.

Among the Pythagoreans, a mystical and religious order founded in the 6th century B.C., the program to determine the rational organization of nature with the help of mathematics became fully realized. The Pythagoreans were fascinated by the fact that phenomena that are physically very different display identical mathematical characteristics, and it became clear to them that mathematical connections must lie behind things, and that these connections must be the essence of the phenomena.

To be specific, the Pythagoreans found this essence in numbers and in the relations among numbers. The numbers were the first principle in the description of nature, and they were the form and matter of the universe itself. The Pythagoreans are supposed to have believed that "all things are numbers." This belief makes more sense in light of the fact that the Pythagoreans represented the numbers as points, which they may have conceived as particles, and that they arranged the points in geometrical patterns, which they may have conceived as representing actual objects. Even if this is speculative, there is nevertheless no doubt that in the course of working to develop and refine their own principles, the Pythagoreans began to grasp the numbers as abstract concepts and physical objects as the concrete realizations of the numbers.

The Pythagoreans "reduced" the motions of the planets to numerical relations by the following considerations. They believed that when the heavenly bodies moved in space, they produced sounds, and that a heavenly body that moves rapidly gives rise to a tone at a higher pitch than one that moves slowly. This is entirely analogous to the sound produced by an object swung around at the end of a string. Further, according to Pythagorean astronomy, the greater the distance of a planet from the Earth, the faster it moved. Therefore, the sounds produced by the planets would vary with their distance from the Earth, and these sounds were all in mutual harmony. Like all harmony, this "music of the spheres," and thus also the motion of the planets, could be reduced to numerical relations.

Other relations in nature could also, according to the Pythagoreans, be reduced to numbers. The numbers 1, 2, 3, and

4, which they called *tetractys*, were particularly valued. Thus, the Pythagorean oath is said to have run, "I swear in the name of the tetractys, which have been bestowed on our soul, that the source and roots of the overwhelming nature are contained therein." Nature was made up of fourfold groupings, such as the four geometrical elements (point, line, surface, and space), or the four material elements emphasized later by Plato (earth, air, fire, and water).

The sum of the four numbers in tetractys is 10, and therefore, the Pythagoreans regarded the number 10 as the ideal number, representing the universe. Since 10 was the ideal number, there had to be 10 heavenly bodies in the universe. Perhaps in order to achieve the "required" number of heavenly bodies, the Pythagoreans introduced a *central fire* around which the Earth, the sun, the moon, and the five planets known at that time revolved, and an *anti-earth* on the other side of the central fire. In the Pythagoreans' model of the universe, the number 10 reappears, too, for the central fire is surrounded by 10 (invisible) concentric spheres, of which the sphere next to the innermost one carries the Earth around the central fire in the course of one day. The innermost carries the anti-earth, which also circles the central fire in the course of one day diametrically opposite the Earth. Neither the central fire nor the anti-earth is visible from the Earth on which we live, since it is turned away from them. Beyond the Earth come the spheres for the heavenly bodies in the following order: moon, sun, Venus, Mercury, Mars, Jupiter, and Saturn. Each of these spheres rotates from west to east in the course of a specific period of time equal to the period of revolution of the heavenly body. Outermost is the 10th sphere, which contains the fixed stars; this sphere moves as well, but so slowly the eye cannot detect it.

By the preceding, the Pythagoreans had constructed an astronomical theory based on numerical relations. Their astronomical theory constituted not mere number mysticism, however, for it appears also as a clear attempt to explain a number of astronomical phenomena with the aid of a mechanical model. Thus, the anti-earth was introduced not only to bring the number of the heavenly bodies up to 10; it was also intended to explain why more lunar eclipses occurred than solar eclipses. Whether by luck or intuitive genius, it thus fell to the lot of the Pythagoreans to set forth the two theses that proved later to be altogether crucial for natural philosophy—first, that nature is based on mathematical principles, and second, that relations among numbers underlie, unite, and explain the order found in nature.

The Greek who most effectively promoted mathematical investigations of nature was Plato (427–348 B.C.). Plato went further than most Pythagoreans in that he did not wish only to understand nature through mathematics, but to reach beyond nature and comprehend the mathematically organized ideal world he believed to be the true reality. The sense-oriented, the impermanent, and the imperfect were to be replaced by the abstract, the permanent, and the perfect. He hoped that a few fundamental observations of the physical world would furnish a basis from which truth could be evolved through reasoning, and when that was achieved, observation would no longer be necessary. From this point then, nature would be totally replaced by mathematics. For Plato, mathematics was not merely the bridge between the ideas and the things we can observe with our senses; mathematical order was the true description of nature. It was also Plato who laid the foundations of the axiomatic-deductive method, which he considered the ideal method for systematizing and gaining new knowledge.

The need to develop mathematics to study the physical world and to achieve true knowledge about it was also emphasized by the most outstanding of the successors of Plato, namely, Aristotle (384–322 B.C.). Though Aristotle and his school differed with Plato at crucial points regarding the connection between mathematics and the real world, this school, too, argued for the mathematical formulation of nature. Aristotle acknowledged that the mathematical abstractions were derived from the material world, but there are no passages in his writings that can be taken as supporting the view of mathematics as a correction to or an extension of the knowledge we perceive through our senses. He believed that the motions of the heavenly bodies were on the whole mathematically formulated, but that the mathematical laws were purely a description of events.

Far-reaching in its significance for scientific research and philosophical thought was Aristotle's idea of *causality*. To explain how one event brings about another, Aristotle and his school established a schema according to which all phenomena can be referred to four types of cause: the *material* cause, the *formal* cause, the *effective* cause, and the *final* cause. To illustrate the four types of causation, imagine a sculptor creating a sculpture. Here, the material cause is the stone from which the sculpture is cut, along with the sculptor's tools; the formal cause is the design of the sculpture in the artist's mind; the effective cause consists of the artist's cutting away some of the stone; and the final cause is the purpose the sculpture is to serve in decorating a room or a building. The final cause—also called the *teleological doctrine*—was for Aristotle the most important, for it is this cause that gives the whole activity meaning. The teleological doctrine was the reigning principle in the explanation of nature up to the breakthrough with

Galileo of the mechanistic conception of nature at the beginning of modern times.

When Alexander the Great (356–323 B.C.) set forth to conquer the world, he transferred the center of the Greek world from Athens to a town in Egypt he immodestly dubbed Alexandria. It was in Alexandria that Euclid (c. 300 B.C.) published the first comprehensive work of mathematical knowledge, the classic "Euclid's Elements," used in geometry instruction up to our own century, and which in large measure set the standard for future mathematical literature. Thus, here we find the first proofs. Euclid also wrote treatises on mechanics, optics, and music in which mathematics forms the core. Mathematics was the ideal version of what the known physical world contained.

In the Alexandrian period, from c. 300 B.C. to c. 600 A.D., the Greeks extended mathematical knowledge immensely. One can cite the major work of Apollonius (c. 262–190 B.C.) on conics; several first-class works of Archimedes in many fields of mathematics and mechanics; works on trigonometry by Hipparchus, Menelaus, and Ptolemy (85–165 A.D.); and later in this period, the arithmetic work of Diophantus. All these works, like Euclid's, offered abstract versions of objects, relationships, and phenomena in the physical world.

2. Greek Astronomy and the Ptolemaic System

We turn now to a closer description of the Greek contributions to astronomy and cosmology. A principal reason the Greeks concerned themselves with astronomy was that they regarded the study of the heavenly bodies as the safe path to wisdom, since the most complex motions occurred in them; at least it seemed so to the human eye. Though present-day mathematical astronomy is not the one handed down from the Greeks, their patterns of thought and conjectures have greatly influenced its formation. The earliest mathematical reasoning and the earliest comprehension of cosmological phenomena are to be found among the Greeks.

Even among the most primitive peoples, we find interest in the heavenly bodies. They have revered the light and warmth from the sun, have been terror-struck by eclipses and full of wonder over the bright light shining from the planets, appearing and disappearing at different times of the year, and over the overwhelming panorama of light from the Milky Way. Before the Greeks took up their study, precise knowledge of these phenomena was limited, however, to familiarity with the periods of revolution of the sun and the moon, and with those times in the year when certain planets became visible and disappeared again. The Egyptians and the Babylonians undertook some observations primarily of the motions of the sun and the moon, partly in relation to their calendar, and partly to monitor the seasons with respect to agriculture. Neither these peoples nor other cultures before the Greeks produced a coherent theory for the motions of the heavenly bodies. They all lacked both the requisite mathematical skills, and, like the Greeks themselves, effective instruments of observation. The apparently complicated behavior of the heavenly bodies in their motions prevented them from detecting any hint of plan, order, or regularity. To them, nature seemed thoroughly mystical.

As noted before, the Greeks believed otherwise. Driven by their thirst for knowledge and by their predilection for rational argument, they were convinced that a closer investigation of nature would unveil the order that lay behind the motions of the heavenly bodies. Many of the Greek astronomers proposed and defended ideas that in the end became part of modern cosmology.

The study of the heavenly bodies began in Miletus in Asia Minor. In the 6th century B.C., this was a flourishing city. Industry and commerce had brought wealth to the city and, therefore, the population lived in comfort. They travelled widely, and brought the wealth of Oriental thought from Egypt and Babylon and other places to the city.

Thales, cited earlier, has the honor of being the first scientist and philosopher in Western tradition. The story goes that he once fell into a ditch while contemplating the stars, and he is also known (on somewhat dubious grounds) for having predicted the solar eclipse of 585 B.C. Thales's successors Anaximander (611–549 B.C.) and Anaximenes (570–480 B.C.), continued to produce theories touching on the fundamental questions of the universe and its structure; but mathematics played no significant role in their speculations. Since they lacked both instruments and an established methodology, these scientists could only guess the nature of the heavenly bodies and their distances from Earth. The other planets were not identified as such, being regarded as essentially the same as stars. In Greek, *planet* means 'wanderer.' Besides, it was believed that the stars were closer to us than the sun and the moon. Despite the obvious shortcomings of their theories, Thales and his school went far beyond what previous civilizations had thought, and it is especially remarkable that, in their theories, about the universe, they had no recourse to gods or demons or other external aids not acceptable to science.

Pythagoras, who lived in the 6th century B.C., is the next major figure in Greek philosophy and science. Born on the island of Samos near Miletus, he came at the age of 50 after a peripatetic life to Crotona in Italy, and there gathered his disciples into a remarkable fraternity that combined scientific studies with religious rituals. The Pythagoreans revolutionized cosmology by freely declaring that the Earth is a sphere. We know from writings of Parmenides (c. 500 B.C.) that Pythagoras himself believed this. The reason for this assertion was very likely at least as much aesthetic as scientific, for the Pythagoreans regarded the sphere as the most beautiful of all solids. Gradually, the conception that the Earth is spherical won general acceptance, though Aristotle, writing in the middle of the 4th century B.C., expressed the need for a better argument for it. The theories put forward by the Pythagoreans were purely speculative and moreover had no great influence on later Greek astronomical thought.

It was the highly irregular motions of the planets that now attracted the attention of the astronomers. It had become clear to the stargazers that Venus and Mercury, in contrast to the three other planets then known, always stayed close to the sun, and therefore could be seen only in the morning and in the evening, and they learned to identify "the morning star" and "the evening star" as one and the same. They also carried out observations, and were especially mystified by the planets' *retrograde* (backwards) *motion*: the remarkable way in which the planets sometimes seem to stop in their normal course across the heavens from west to east, linger a while, then wander backwards in a westerly direction a short distance, stop once more, and at last resume their motion eastwards. This confusing behavior became the riddle of the astronomers, embarrassing the Greeks, who believed in order and regularity.

In his academy, Plato took up the challenge and sought a coherent theory of the motions of the heavenly bodies, which, among other things, would reconcile this apparent irregularity.

An answer to the problem was given by Eudoxos, a pupil of Plato. Eudoxus was one of the ablest Greek mathematicians; his theory is the first comprehensive astronomical theory known, and it represents a large step forward in the quest to explain nature rationally.

Eudoxus (c. 408–355 B.C.) came from Cnidus on the west coast of Turkey. As a young man, he traveled to Italy and Sicily to study geometry, and he was widely recognized for his mathematical knowledge. At 22, he traveled to Athens, where he attended Plato's lectures at the Academy. He also carried out his own observations, and centuries later his observatory would routinely be pointed out to curious travelers.

The model of the universe proposed by Eudoxus encompassed a series of concentric spheres, within whose center was the immovable Earth. To reconcile the complicated motions of heavenly bodies set apart from the immovable Earth, Eudoxus assumed that a combination of motions on spherical surfaces would produce the desired orbits of the heavenly bodies. The schema is rather detailed, for each individual body required a system with three or four spheres, and the system varied from planet to planet and for the sun and the moon. The spheres were of course purely mathematical and hypothetical. Eudoxus seems to have been satisfied with these considerations, and probably regarded the whole system exclusively as an elegant theory, which neither made demands on nor required physical verification. Eudoxus's work has been lost, and his design is known only through other writers, Aristotle in particular. It appears that Eudoxus's theory was able in its fashion to describe rather precisely all motions except the orbits of Venus and Mars. For these two heavenly bodies, however, there were serious deficiencies. What the critics especially focused on was that if things were as Eudoxus said, namely, that the heavenly bodies always were at the same distance from the Earth, then they ought not to display variations in brightness or size; that they do so is after all visible to the naked eye. Eudoxus was apparently aware of this himself, but disregarded these difficulties. He and his contemporaries undoubtedly recognized that the refutation of his theory might remove the Earth from the center of the universe, a somewhat troubling outcome for his time.

In the middle of the 5th century B.C., Heraclitus proposed two quite revolutionary ideas. The first is that the Earth rotates about its own axis. This was a daring idea, for it went against the general thinking, and was contrary to the impression of the senses. That the Earth should rotate rather than the universe seemed more rational to Heraclitus based on considerations of size. Even though it sounded reasonable, however, the idea was not immediately accepted.

The other idea was more far-reaching, for it attacked the existing cosmology at one of its most vulnerable points, namely, the inability of Eudoxus's system of concentric spheres to explain the observed variations in size and brightness of the heavenly bodies. Heraclitus pointed the way to possible alternatives. Based perhaps on the fact that Venus and Mercury are always near the sun, he proposed that these two planets move in circles whose center is the sun. If, as Heraclitus said, this heliocentric motion occurs in connection with the sun's own circular motion around the Earth, then the distances from the Earth to Venus and Mercury will obviously vary, occasioning some variation in brightness. This new idea had an immense and lasting influence. On the purely mathematical side, Heraclitus had created for the first time in cosmology the concept of an *epicycle*, that is, a circle whose center moves around on another circle. Much of the later development stems from this beginning. His theory was weakened by its restriction to two of the five planets known at the time, a limitation Heraclitus apparently did not try to remove; but the idea of the sun as a center for heavenly motions was nonetheless a milestone on the road away from the conception of the Earth as the center of the universe.

The interest in quantitative knowledge and the willingness to gather such knowledge appeared in the second great Greek period, during which the center of civilization had moved to Alexandria. In this city, the Greeks came into closer contact with Egyptians and Babylonians, and they had easier access to the astronomical observations made by these peoples for millennia. It was also significant that Alexander's successors in Egypt instituted a great center for scholars, which they called the Museum, and contributed lavishly to the creation of a renowned library there. During the Alexandrian period, Eratosthenes, Apollonius, Aristarchus, Hipparchus, Ptolemy, and many others devoted themselves to studies of geography and astronomy here. The Alexandrians produced the astronomical theory that came to dominate for more than 1,500 years.

Prominent among the contributions of the Alexandrians was the heliocentric hypothesis, advanced by Aristarchus of Samos (310–230 B.C.). Early in his career, Aristarchus made the first known attempts at calculating the magnitudes of the heavenly bodies and the distances among them. From a modern perspective, this work is an exercise in simple trigonometry; his work, however, was prior to the discovery of trigonometry, and he was content to determine the upper and lower limits of such magnitudes, rather than their precise value. His main tool was the geometrical synthesis carried out a generation or two before by Euclid. Comparing Aristarchus's results with modern estimates, one could well say that they were distinctly erroneous. The error was not in his mathematics, however, but in the observations possible with the primitive instruments of that time; but Aristarchus had taken the first steps on the road to future conquests by asking, "How far?" and "How big?"

Nowhere in the single surviving work of Aristarchus does he maintain that the Earth moves around the sun, but we know that he had that conception from a writing of Archimedes, who was his junior by just 23 years. We cannot be certain about Aristarchus's motives, but Heraclitus after all had proposed a useful idea in having Venus and Mercury orbiting the sun. This idea, combined with Aristarchus's own conception of magnitudes, and with a certain intuition of the principles of dynamics, might have convinced him that it would be physically more reasonable for the smaller body to move around the larger. Perhaps, too, he regarded the heliocentric model simply as an attractive hypothesis worth pursuing for the sake of its mathematical consequences. In any case, however, the idea was too advanced for his time and attracted little support.

Immediately following the pioneering work of Aristarchus in measuring the magnitudes and distances of the heavenly bodies, another brilliant scientist, Eratosthenes, who was born in or about the year 276 B.C. in Cyrene (Aswan) in North Africa, attempted to measure the circumference of the Earth. Eratosthenes observed that at noon on the summer solstice, the sun cast no shadow in Cyrene, while at the same time in Alexandria it cast a shadow on 1/50 of a complete circle. Since he assumed that the two cities lay on the same meridian, and that the rays of sunlight, arriving at different places on the Earth, are parallel, he could by an easy geometrical argument (Fig. 4.1) demonstrate that the distance along the surface of the Earth between Alexandria and Cyrene is 1/50 of the Earth's circumference. Eratosthenes was mistaken in placing Cyrene

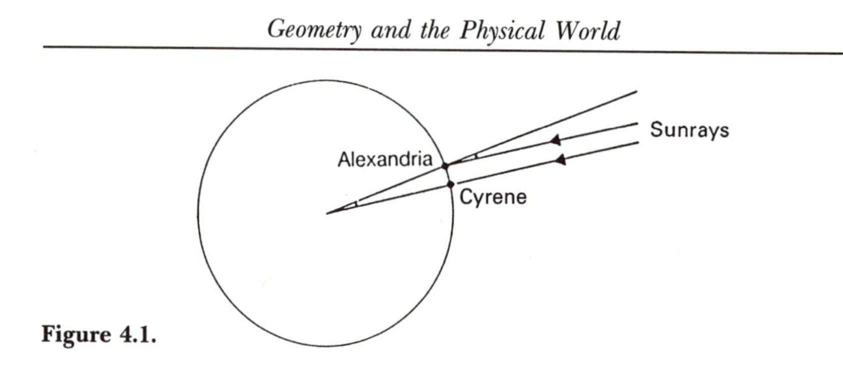

Figure 4.1.

and Alexandria on the same degree of longitude, but the significance of his contribution stems less from its degree of precision than from the example it set. Here was a stimulus to the continued building of a quantitative cosmology.

The zenith of Greek astronomy was reached with Hipparchus (who died c. 125 B.C.), and especially Ptolemy (who died in 168 A.D.).

It was clear to Hipparchus that Eudoxus's theory, which assumed that the heavenly bodies were tied to rotating spheres whose center was the center of the Earth, could not account for many facts observed by Hipparchus himself and by other Greeks. Instead of Eudoxus's schema of explanation, Hipparchus assumed that a planet P moved around a circle, the epicycle, with a constant velocity, and that the center Q of this circle likewise moved with constant velocity, but around another circle whose center was the center of the Earth. See Fig. 4.2. With appropriate choices of radii of the two circles and the velocities of P and Q, he was able to achieve a quite precise description of the motions of many planets. The motion of a planet according to Hipparchus's schema corresponds to the motion of the moon according to modern astronomy. The moon travels around the Earth, which in turn travels around

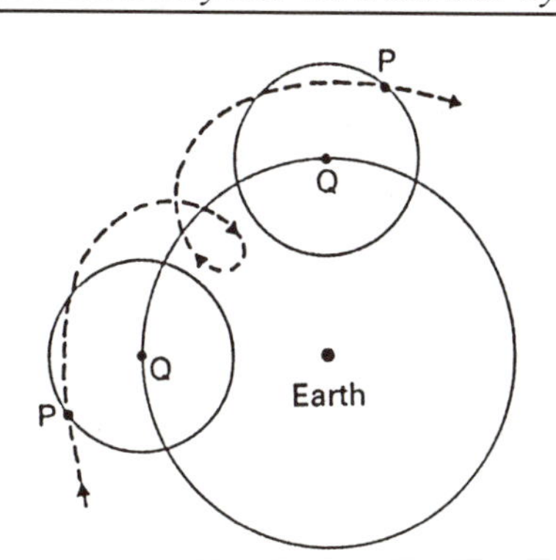

Figure 4.2.

the sun. The motion of the moon around the sun thus resembles the motion of a planet around the Earth in the system of Hipparchus.

For some heavenly bodies, Hipparchus found it necessary to employ three or four circles that moved around each other. A planet P thus moves in a circle around the mathematical point Q, while Q moves in a circle around the point R, and R moves around the Earth, every point or object moving at its own velocity. See Fig. 4.3. In some cases, Hipparchus had to assume that the center of the innermost circle, called the *excenter*, did not lie precisely at, but only close by, the center of the Earth. Motion corresponding to this last geometrical construction was called *excentric*, whereas it was called *epicyclic* if the center lay at the center of the Earth. By using both types of motion and by appropriate choices of radii and velocities for the circles in question, Hipparchus was in a position to describe the motion of the moon, the sun, and the five planets known at his time quite well. With this theory, lunar eclipses could be predicted to within an hour or two, while predictions of solar eclipses were somewhat less precise.

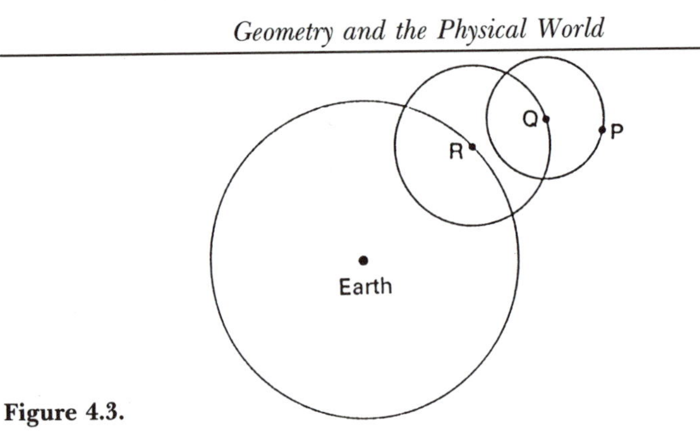

Figure 4.3.

It is worth noting that from a modern viewpoint, Hipparchus was actually taking a step backwards, since about a century earlier Aristarchus had—as we know—proposed the theory that all planets revolve about the sun; but observations carried out over a 150-year period at the observatory in Alexandria together with earlier Babylonian observations convinced Hipparchus of what we today know is the case, namely, that a heliocentric theory in which the planets move in "circles" around the sun is not viable. Instead of attempting to refine Aristarchus's idea, Hipparchus discarded it as overly speculative.

In the 2nd century A.D., we find the truly greatest figure in Greek astronomy, Ptolemy, who in his work *Syntaxis Mathematica*, better known as the *Almagest* (the name given it by Arabic translators), presented a comprehensive overview of Greek astronomy. Hipparchus's work, for example, is known only through Ptolemy. The *Almagest* consists of 13 books, and the work dominated European astronomy for 1,400 years.

Mathematically, the *Almagest* brought Greek trigonometry to the definitive form, which it maintained for more than 1,000 years. Astronomically, it gave an original presentation of the geocentric theory with epicycles and excentricity that has become known as the *Ptolemaic system*. This theory was quantitatively so precise, and was generally accepted for so long, that people were deceived into regarding it as absolute truth. The theory is the ultimate Greek response to Plato's problem of explaining the motions of the heavenly bodies rationally, and it was the first truly great scientific synthesis. With Ptolemy's completion of Hipparchus's work, the mathematical model fit the known universe completely "down to the 10th decimal place." Ptolemy, however, clearly conceived his theory as only a mathematical construct.

Ptolemy was aware of Aristarchus's heliocentric theory, but dismissed it on the grounds that if the Earth were in motion, it would leave lighter objects such as animals and human beings behind. His astronomy takes as its point of departure that the heavenly bodies are spherical. He regards this as the oldest cosmological truth known to mankind, but substantiates it by and large on the basis of observations. Ptolemy felt it necessary also to offer proofs of an observational nature that the Earth was spherical. As mentioned, he insisted that the globe of the Earth does not move, although he seems to have been aware that its rotation would produce some of the phenomena we observe in the heavens. The Earth is in the middle of the universe, and compared with its distance to the stars, its size is no more than a point, says Ptolemy.

In the third book of the *Almagest*, Ptolemy takes up the problem of the sun's orbit, proposing essentially the model Hipparchus suggested, namely, that the center of the sun's circular motion lies near, but not at, the Earth's center. He assumed this position to obtain as simple an explanation as possible, observing that it is simpler to solve the problem by moving one body than two. In his theory of the moon, Ptolemy finds that Hipparchus's model (with epicycles and excentricity) fits the observations of the relations of magnitude of the new moon and the full moon, but breaks down in the intermediate phases. To solve this problem, Ptolemy had the moon's excenter rotating around a little circle of its own, whereby the moon is periodically drawn closer to the observer. This adjusted model gives the variations in the size of the moon with great accuracy.

In the ninth book of the *Almagest*, we find Ptolemy's unique major contribution, namely, the first thoroughly elaborated and comprehensive account of the irregular motions of the planets. As a first approximation, Ptolemy refers all the planetary motions to the *plane of the ecliptic* (the plane of the sun's circular orbit), which he imagines undergoing a slow rotation, like a spinning top just prior to toppling. This rotation produces *precession*, that is, the annual advance of the equinoxes. Further, he finds the simple schema of epicycles and excentricity insufficient to describe the orbits of the planets, since this schema, contrary to observation, implies that the retrograde arcs are equal in length and uniformly distributed in space. Ptolemy gets around this problem by postulating that each planet's epicycle moves uniformly not with respect to its excenter, but relative to another point, called the *equant*. The planet moves around the epicycle in the same direction that the center of the epicycle moves around the excenter; retrograde motion arises when the planet is on the side of the epicycle nearest the Earth. Only Mercury requires yet another complication, which corresponds to the device Ptolemy contrived for the moon: Mercury's excenter travels about a little circle of its own, and thus the planet's epicycle is drawn periodically nearer the Earth. An "inner" planet, Mercury or Venus, runs through its epicycle during its own year (equivalent to its period of revolution around the sun), while the epicycle's motion around the excenter takes an Earth year. For the "outer" planets, the arrangement is reversed—the time of motion of the epicycle around a planet's excenter is equivalent to what we will now describe as the planet's own period of revolution around the sun, whereas the motion along the epicycle corresponds to what for us would be the Earth's period of revolution around the sun. Each epicycle is slanted relative to its excenter-circle such that one obtains, by and large, that the epicycle is parallel to the ecliptic.

The idea of the equant must be regarded as Ptolemy's masterpiece. It describes the observed phenomena very effectively, and is the forerunner of Kepler's ellipses. Some of Ptolemy's later critics say that the introduction of the equant seems to compromise the sacred first principle of motion of a heavenly body, which insists on uniformity only with respect to the center of a circle. The equant became, in the eyes of some, a veritable scandal, which among other things convinced Copernicus that it must be the Earth and not the sun that was in motion.

Within its self-imposed restriction to even circular motions, the Ptolemaic system described the motions of the heavenly bodies with a precision that answered fully to the underlying observations. The *Almagest* must be numbered among the most influential works in the history of science, even though many aspects of its presentation, especially its belief in a central, immovable Earth, have proven not to hold true.

Regarding the pursuit of truth, it is worth noting that Ptolemy understood fully that his theory was only a useful mathematical description fitting the observations, and not necessarily a true description of the arrangement of nature.

Faced with the choice between alternative models for some of the planets, he always chose the mathematically simplest ones.

3. The Copernican World, Tycho Brahe, and Kepler

Greek civilization was destroyed by the conquests of the Romans and the Mohammedans, and with its demise, southern Europe entered the Middle Ages, which lasted there from about 500 to 1500 A.D. In this period, culture was dominated by the Catholic Church, which regarded life on Earth as a preparation for life after death in heaven. Therefore, studies of physics and astronomy—mathematical or otherwise—were not encouraged. Nonetheless, there were of course individuals and groups of individuals who made efforts to continue mathematical and physical investigations. In particular, they used mathematics to verify physical phenomena, and some insisted also on experimental techniques. European thinkers in the Middle Ages believed that the universe was fundamentally rationally ordered, and that mathematical thinking could produce knowledge of it. Nor should one overlook the contributions to mathematical knowledge made by the Hindus and Arabs during the medieval period, alluded to in connection with ornamentation in Chapter 1.

The modern period, with which we shall largely occupy ourselves here, can be said to have begun around the year 1500.

The 16th century is often extolled as the Renaissance, when Greek thought and ideas were reborn. The Europeans did not immediately embrace the whole spirit of Greek thought, however. The period is often characterized as *humanistic*, as studies of the Greek works were more prominent than active pursuit of Greek scientific ideals, which were, after all, to employ reason to study nature and seek its underlying mathematical structure. At one point, there was a serious problem, since Greek ideals conflicted with the reigning culture of the 1500s. Where the Greeks believed nature was consistent with an ideal mathematical plan, medieval thinkers referred the whole plan and its realization to the Christian god. He was the architect and creator, and all events in nature followed the plan He had established. The universe was God's work, subordinate to His will. The mathematicians and scientists of the Renaissance and several centuries after were orthodox Christians, accepting the Catholic teachings that in no way embraced the Greek idea of a mathematically structured nature.

How could the attempts to understand God's universe be reconciled with the search for mathematical laws of nature? Well, the problem was solved by adding a new doctrine, namely, that the Christian god itself had structured the universe mathematically. The Catholic doctrine, which placed the greatest importance on searching out and understanding God's will and creations, thus took the form of an investigation of God's mathematical arrangement of nature. In reality, mathematical works in the 16th, 17th, and most of the 18th centuries were religious treatises whose authors sought to discover the mathematical laws of nature in devotion to God for the magnificence of his creations. Of course, human beings could not hope to comprehend all of the divine plan as clearly as God himself understood it, but they could at least, with humility and modesty, come nearer to God's intentions, and in this way understand God's world.

One can go even further and assert that these mathematicians were altogether certain that mathematical laws were at the foundations of natural phenomena. They insisted on seeking these laws, for they were convinced *a priori* that God had incorporated them in His construction of the universe. Each discovery of a law of nature was construed as an expression of God's genius, rather than genius in the person who had brought it to light. The mathematicians and scientists of Renaissance Europe worked on the basis of deep-rooted Christian faith. Thus, the recently rediscovered Greek works confronted a profoundly Christian view of the world, so that those intellectual leaders who had grown up in the one belief system, but felt drawn to the other, combined medieval Christianity with Greek science.

Together with this forceful new intellectual eclecticism, the idea of "return to nature" began to gain support. All kinds of scientific people abandoned the endless speculations based on dogmatic principles, which made little sense and were unrelated to the world of experience. They turned instead towards nature itself as a reliable source of knowledge. What is certain is that around 1600, the Europeans were motivated to carry out what has been described as the scientific revolution. Various events accelerated these developments: geographical explorations and discoveries of new lands and peoples; the invention of the telescope and the microscope, which revealed new phenomena; the compass, which aided navigation; the heliocentric theory, reintroduced by Copernicus, which stimulated new thinking about our planetary system; and finally, the Protestant revolution, challenging Catholic doctrine. Soon thereafter, mathematics assumed its leading role as the key to physics and astronomy.

As mentioned, in the Middle Ages the Catholic Church maintained that the Earth was the center of the universe. It was Copernicus (1473–1543) who first seriously led thinking in a new direction. In the time of Copernicus, the Arabs, in their efforts to improve the precision of the Ptolemaic system, had added several epicycles to the model; their theory required, all in all, 77 circles to describe the motions of the sun, the moon, and the five planets then known. For many astronomers, among them Copernicus, this theory had become ridiculously complicated. Harmony in nature required a simpler theory than the complex elaboration of the Ptolemaic system.

Copernicus, who had read that some of the Greeks, Aristarchus in particular, had proposed that it was the Earth that moved around the fixed sun, rotating simultaneously on its own axis, decided to investigate this possibility. In a sense, he was overly impressed by Greek thought, for he also believed that motions of the heavenly bodies had to be circular, or at least a combination of circular motions, since circular motion was the "natural" motion. Copernicus also accepted the Greeks' belief that every planet had to move at a constant velocity around its epicycle, and that the center of each epicycle had to move at a constant velocity around that circle on which it was set. Such principles were axiomatic for him. Into the bargain he added an argument that reveals the bend towards mysticism among the thinkers of the 1600s. He found that a variable speed could be caused only by a variable force, but God, the first cause of all motion, is constant.

The outcome of all Copernicus's reasoning was that he employed the schema with epicycles and excentricity to describe the motions of the heavenly bodies. The crucial difference in his version however, was, that now the Earth was no longer the center of the universe. Instead, he placed the sun in this position. Hence, the Earth moved around the sun like the other planets, in a circular orbit. Simultaneously, the Earth rotated around its own axis. Finally, he demonstrated that all the complicated relations among the motions of the planets were only apparent. The appearance they gave of now and then standing still or going backwards was due entirely to our observing them from the Earth, which was itself in motion. When the Earth was moving straight towards a planet, the planet had to appear to stand still. If the Earth overtook a planet, the planet would seem to fall behind, that is, seem to be traveling backwards. Copernicus was unable to explain, on the other hand, why the planets moved around the sun, just as he was unable with any precision to indicate the laws determining their motions.

With his model, Copernicus was in a position to reduce the number of circles required to describe all the motions of the planets from 77 to 34. The heliocentric viewpoint thus permitted significant simplifications in the description of planetary motions, but otherwise the model was not overwhelmingly accurate. He tried variations, therefore, in which the sun was, of course, always stationary and either placed in or nearby the excenter of each of the planets. Even with these variations he failed to achieve much greater precision, but this did not dampen his enthusiasm for the heliocentric viewpoint.

The heliocentric theory, which de-emphasized the importance of the human being in the universe, naturally encountered stiff resistance. Thus, Martin Luther called Copernicus an "amateur astrologer," and "a fool who wants to stand astronomical science on its head." The Inquisition condemned the new theory as "this false Pythagorean doctrine, which strongly conflicts with the Holy Writ." In an official message, the Catholic Church called the Copernican theory, "more odious and more harmful to Christendom than anything whatsoever contained in Calvin's and Luther's writings or in all other heresies."

The Copernican system broke through only slowly. Thus, the great Danish astronomer Tycho Brahe (1546–1601) did not recognize the Copernican system, and sought to establish his own, the *Tychonic system of the world*, in which the moon and the sun revolve around the Earth, while the other planets, as in the Copernican system, revolve around the sun. Tycho thought, namely, that since he could not detect any sign of its movement, the Earth had to be stationary. Tycho's system, however, attained no real significance, and it was—as we know—Copernicus who carried the day. During his long sojourn on the island of Hven, Tycho Brahe had made a great many extremely precise observations of the planets, especially of the position of Mars in the sky, and when he abandoned Denmark (after a strong disagreement with King Christian IV), he brought all his records with him to Prague. After his death in 1601, these were taken over by his collaborator Johannes Kepler (1571–1630) to be interpreted. Through many years' research, this gifted scientist succeeded in discovering the laws of motion of the planets around the sun.

Kepler, too, exhibited a mystical turn of mind, believing, like Copernicus, that God's arrangement of the universe was consistent with several simple and elegant mathematical schemas, and that all of nature and the sparkling sky were symbolized by geometry; but Kepler had other qualities as well that we associate today with a serious scientist. Although his inventive mind led to the conception of many new theoretical hypotheses, he knew that theories have to fit observations, and especially in his later years he realized that empirical data can also reveal fundamental principles in nature. Therefore, he relinquished his beloved mathematical hypotheses when he

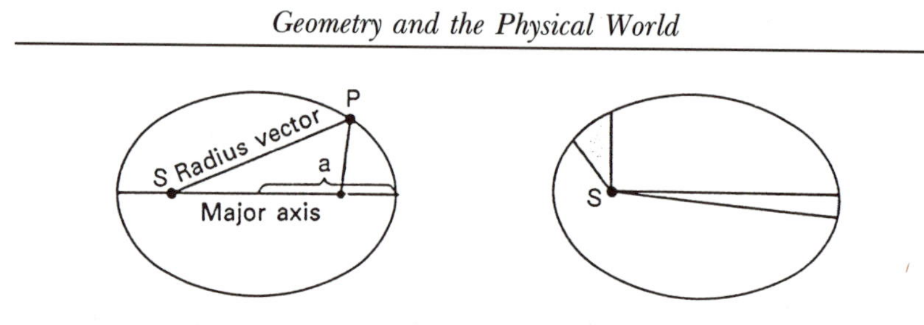

Figure 4.4. **Figure 4.5.**

found they did not fit the observations, and it was precisely this consistent refusal to accept whatever deviated from empirical data that led him to the crucial new ideas. He also possessed the humility, the patience, and the energy required to carry out pioneering work. The result was the set of three famous Keplerean laws:

(1) The planets move around the sun in elliptical orbits, with the sun at one of the two foci.

(2) During the motion of a planet, areas swept out by the radius vector within equal time intervals are equal.

(3) The squares of the planets' periods of revolution around the sun vary as the cubes of their mean distances from the sun.

These three laws warrant closer examination.

The first Keplerean law is purely geometrical. An ellipse is a closed plane curve with the characteristic that the sum of the distances from all its points to two fixed points, called the *foci*, has constant value. See Fig. 4.4. The sun is always located at one of the foci, e.g., as in the figure at S. The line from the sun to the planet, designated P in the figure, is called the *radius vector*. Half the orbit's major axis (the line through the foci) is called the *mean distance*, and is designated a. By the orbit's excentricity, we mean that fraction of the semi-major axis constituted by the sun's distance from the orbit's center.

A planet's velocity varies according to the second Keplerean law, as illustrated in Fig. 4.5. Thus, a planet's speed is highest when it is nearest the sun, and lowest when it is farthest away from the sun. At the beginning of January, the Earth is nearest the sun, while it is farthest at midsummer. Therefore, the sun travels faster across the sky at the beginning of January than at midsummer. The differences, however, are not great, since none of the planetary orbits deviate very much from the circular one. The Earth's orbital excentricity is thus only 1/60. That we actually are nearer the sun in the winter than in the summer can be corroborated by measuring the diameter of the sun, which really does seem a bit larger in the winter than in the summer.

If we measure a planet's period of revolution T in Earth years, and its mean distance a relative to Earth's mean distance from the sun, then Kepler's third law states that the period of revolution multiplied by itself is equal to the mean distance multiplied by itself twice, or as a mathematical formula, that $T \cdot T = a \cdot a \cdot a$ (in short form, $T^2 = a^3$).

With the help of the Keplerean laws, we can calculate in advance a planet's position in its orbit at any time. Copernicus had said that to explain a phenomenon, one must proceed from the simplest hypotheses fitting the facts. Copernicus used exactly this argument against the Ptolemaic theory, and since he believed that the universe was God's creation, he conceived the simplicity he had found to be the true structure of the universe. Since Kepler's mathematics was simpler still than Copernicus's, he had all the more reason to believe he had found the very laws that God had incorporated in the construction of the universe. There were important scientific arguments against the notion that it was the Earth that moved, and the theory came up against religious and philosophical conservatism. Despite this, the new theory gradually won acceptance, for especially after Kepler's work, mathematicians and astronomers were impressed by its simplicity. The theory was also much more useful in navigational calculations and in the construction of calendars, so that many geographers and astronomers began to make use of it whether or not they were convinced of its truth.

The heliocentric theory found an exceptionally able advocate in Galileo Galilei (1564–1642). Born in Florence, Italy, he entered the university in Pisa at the age of 17 to study medicine. Reading the works of Euclid and Archimedes, however, his interest in mathematics and science was awakened, and thus he turned to these fields.

In the summer of 1609, Galileo heard of a Dutch invention that made it possible to see distant objects as clearly as objects close by. Galileo was not late constructing a telescope of his own. Gradually, he improved his lenses and ultimately achieved a magnifying power of 30. At a dramatic demonstration for the Venetian Senate, Galileo proved the power of his telescope by spotting enemy warships approaching Venice almost two hours before their arrival.

Galileo had greater plans for his instrument, however. By aiming his telescope at the moon, he observed deep craters and impressive mountains, and he was able thereby to do away with the conception that the surface of the moon was smooth. By observing the sun, in 1610 he discovered sunspots and solar flares. (From these observations of the sun, Galileo suffered serious eye damage, which eventually rendered him almost blind.) He also discovered that Jupiter had four moons orbiting it. (We now know it has at least 16.) This discovery proved that a planet, like the Earth, can have satellites. Copernicus had predicted that if human vision could be enhanced, one would be in a position to observe phases of Venus and Mercury, that is, observe that larger or smaller portions of the hemisphere of the planet facing the Earth are illuminated by the sun—in the same way, one can observe phases of the moon with the naked eye. Galileo found indeed that Venus displayed phases. Here was further documentation of the fact that the planets were of the same character as the Earth, and that they certainly were not perfect bodies composed of one ethereal substance or another, as the Greek and medieval thinkers had believed. With the telescope, it was possible to establish that the Milky Way was not just a broad band of light, but that it consisted of thousands of stars, each of which radiated light. There were other suns and most likely also other planetary systems in the universe. Additionally, it was clear that the universe contained more than seven bodies in motion—a number that gradually had become accepted as sacrosanct. These observations convinced Galileo that the heliocentric theory was the correct one.

Since the Church was displeased with Galileo's championing heliocentrism, the Roman Inquisition in 1616 declared the doctrine heretical and condemned it. Despite this condemnation of works on Copernicanism, Pope Urban VIII in 1620, believing there was no danger that the new theory could ever actually be proved true, nonetheless granted Galileo permission to publish a book on the subject. By virtue of this permission, Galileo was able to compare the geocentric and heliocentric theories in his work, *Dialogo Sopra i Due Massimi Sistemi Del Mondo* (*Dialogue concerning the Two Major Systems of the Universe*), published in 1632. In an effort to please the Church, and thereby escape censure, he included a foreword in which he said that the heliocentric theory was merely a figment of the imagination. Though Galileo had presented the geocentric and heliocentric theories as equally valid, his own preference for the latter was apparent. Galileo wrote so effectively that the Pope began to fear the arguments for the heliocentric theory would damage Catholic faith severely after all. Therefore, Galileo was called again before the Roman Inquisition, and forced under threats of torture to declare, "The falsehood of the Copernican system cannot be doubted, especially not by us Catholics." In 1663, Galileo's work was placed on the list of banned books, a condemnation not lifted until 1822.

In the space age, no one doubts the truth of the heliocentric theory any longer. In the 17th and 18th centuries, however, it was quite possible for people to remain skeptical, since the absence of any sensory experience of a rotating Earth circling the sun argued against the theory. Yes, even those in a position to understand the writings of Copernicus, Kepler, and Galileo had reasons for skepticism, since the mathematical arguments of Copernicus and Kepler were not substantial, so that the whole theory actually rested on a deeply philosophical trust in the simplicity of the heliocentric model.

4. The Breakthrough of Modern Natural Science

From about 1600 onwards, European scientists were undoubtedly convinced of the importance of mathematics in the study of nature. This is apparent in the willingness displayed by Copernicus and Kepler to reject religious doctrines in favor of an astronomical theory that in their time had only mathematical advantages.

The amazing success of modern science and its close relationship with the development of new mathematics in the

17th and later centuries would not have come about had the footsteps of the past been followed. In the 17th century, however, we find two great figures: René Descartes (1596–1650) and Galileo Galilei. These "giants," as Isaac Newton later called them, reformed and reformulated the very method by which scientific activity was pursued. They selected the concepts science would employ, redefined the aims of scientific activity, and transformed scientific methodology itself. This reformulation brought not only a previously unknown vigor to science, but also tied natural philosophy inextricably to mathematics. Practically speaking, their program reduced theoretical physics to mathematics.

Descartes was quite explicit in his belief that mathematics was the essence of physics. He once wrote that he "neither allows for nor hopes for principles in physics other than those that lie hidden in geometry or in abstract mathematics, for in this way all phenomena of nature will yield to explanation, and a deduction of them can be given." The objective world is tied to space and inherent in geometry, and therefore its characteristics must be deducible from first principles of geometry. Descartes insisted that the most fundamental characteristics of a substance (matter) are its shape, its extension in space, and its motion in space and time. Since shape is itself merely a question of extension in space, it must be extension and motion that constitute the basic foundation. Thus, Descartes has said, "Give me extension and motion and I shall construct the universe."

Although Descartes glorified the mathematical method and was completely convinced that he could reduce all science to mathematics, he employed surprisingly little mathematics himself. Apart from results communicated in letters, he wrote only one short book on mathematics, the famous *La Géométrie*, in which, independently of Fermat, he created analytic geometry.

This book was one of three appendices to his great philosophic work, *Essais Philosophiques*, from 1637. Here, Descartes created a general and systematic philosophy, and by reducing natural phenomena to purely physical events, he did much towards freeing science from mysticism and forces of the occult. In addition, by tackling almost all the scientific problems of his time with fresh new concepts and methods, he stimulated others to produce new theories.

Descartes' writings were extremely influential in the second half of the 17th century, and his deductive and systematic philosophy impressed his contemporaries. Newton especially was strongly affected by the significance that Descartes attributed to motion. The superiority of human reason, the invariance of the laws of nature, the doctrine of extension and motion as the essential elements of physical objects, the distinction between soul and body, and the distinction between qualities that are real and inherent in objects and those only apparently present, due in fact to the brain's response to sensory data—all these things are developed in Descartes's writings and have been significant in the formation of modern thought.

Also, Galileo produced a natural philosophy. In broad outline it agrees with Descartes's philosophy, but from the standpoint of science it proved to be a more effective philosophy at certain crucial points. Galileo's great scheme for reading the book of nature included a completely new definition of the aims of science and of the role mathematics was to play in attaining them. It is Galileo's schema of studying and mastering nature that created modern mathematical physics. What led Galileo to his revolutionary ideas about the methodology of science is unclear. He knew that Ptolemy had conceived his own geocentric theory exclusively as a mathematical schema, and that Copernicus had stressed the simplicity of the mathematics in his defense of the heliocentric theory. Kepler did the same, but Galileo disregarded Kepler's work. Galileo agreed with both Copernicus and Ptolemy—the world had been created mathematically. In the preface of this book, we cite a passage from his famous little paper, *Sidereus Nuncius* (*The Starry Messenger*) from 1610, in which he says that nature is written in a great book in the language of mathematics, for nature is simple and well-ordered, and its behavior regular and necessary. Nature acts in agreement with perfect and unchanging mathematical laws. God laid down in the world a powerful mathematical necessity which men—though their reason is related to God's—can grasp only through hard work. Mathematical knowledge is therefore not only the absolute truth, but as sacred as any passage in the holy scripture. Yes, mathematical knowledge is in fact superior to scripture, for while there are disagreements concerning the content of the latter, there is no inconsistency in mathematical truth.

Although Descartes had taken a step towards finding the laws of motion, he did not really address the problems created by the introduction of the heliocentric theory. According to this theory, the Earth rotated simultaneously around its own axis and around the sun. If this was the case, why should objects remain on the Earth? Why did objects fall towards the Earth if the Earth was not the center of the universe? Furthermore, all motions—of projectiles, for instance—appeared to occur as though the Earth was at rest. There was a need for new principles of motion to make sense of these phenomena.

The revolutionary new scheme proposed by Galileo, and pursued by his followers, aims at obtaining quantitative descriptions of scientific phenomena, independent of the possibility of obtaining physical explanations. Galileo's scheme can be understood better by an example. Given the simple situation in which a ball is dropped from a tower (the Leaning Tower of Pisa, say), one might speculate endlessly on *why* the ball falls. Galileo advised a different approach. The distance the ball falls from its starting point increases with time from the moment it is dropped. In mathematical terms, the distance the ball falls, and the time that passes during the fall, are called *variables*, since each of these quantities undergoes a change as the ball falls from different heights to the surface of the Earth. Galileo investigated whether there was a mathematical relation between these variables. The answer he obtained can be written in a simple mathematical formula. If the distance of the fall is d and the duration of the fall is time t, then the formula is $d = \frac{1}{2} \cdot g \cdot t^2$, where g is a constant, depending on the degree of latitude. In Denmark (latitude 55°), $g = 9{,}815$, when d is measured in meters and t in seconds. The formula states that the number of meters d the ball falls in t seconds is 4.9 times the square of the number of seconds. For example, the ball will fall 44.1 meters in 3 seconds, and 78.4 meters in 4 seconds, and so on. We note that the formula is concise, precise, and quantitatively complete. For every value of the one variable, time t in this case, one can calculate precisely the value of the other variable, here the distance d. This calculation can be undertaken for millions of values of time, or really an infinite number of values. The simple formula $d = \frac{1}{2} \cdot g \cdot t^2$ thus contains an infinite amount of information.

It is important to make a distinction here, however. The mathematical formula is a description of what happens; it is not an explanation of a causal relation. The formula $d = \frac{1}{2} \cdot g \cdot t^2$ says nothing about *why* a ball falls; it merely gives quantitative information about how a ball falls. Furthermore, although formulas are used to relate variables that scientists assume are causally related, it is no less true that one need not investigate or understand the causal connection in order to deal with the situation successfully. It was this fact that Galileo so clearly saw as he emphasized mathematical description over the less successful qualitative and causal questions about nature that absorbed the Greek philosophers so deeply—perhaps Aristotle most of all.

On this foundation, Galileo resolved to seek mathematical formulas that describe the behavior of nature. There seems to be no real value at first in such bare mathematical formulas, for explaining nothing, they only describe unvarying regularity in a precise language. Nevertheless, one can say that such formulas have proven extraordinarily fruitful in the exploration of nature. The amazing practical as well as theoretical conquests of modern science have been achieved primarily through quantitative descriptive knowledge, more than through metaphysical, teleological, or even mechanistic explanations of the causes of phenomena. The most important of Galileo's works for posterity is *Discorsi e Dimonstrazioni Matematiche intorno a Due Nuove Scienze* of 1638, which contains his mechanistic investigations. Here he declares, inter alia, that "The cause of the acceleration of a falling body is not a necessary part of the investigation." More generally, he says he is going to investigate and demonstrate some of the characteristics of motion without reference to what its causes might be. Positive scientific questions must be distinguished from questions regarding the ultimate cause, and one can forgo speculations regarding physical causes.

One's first reaction to Galileo's ideas is probably negative, for descriptions of phenomena by means of formulas seem hardly more than a first step. It might appear that the true function of science is rather to be found with Aristotle, namely, in attempting to explain physically why phenomena occur. Even Descartes protested Galileo's resolve to seek descriptive formulas, declaring in one place, "Everything Galileo says about bodies falling in empty space is built without foundation; he ought first to have determined the nature of weight." Furthermore, Descartes says, Galileo should consider ultimate causes. In light of the subsequent development, however, we can see now that Galileo's decision to strive for description was the deepest and the most fruitful idea anyone has ever had about the methodology of the exact sciences.

The decision to seek those formulas that describe a phenomenon leads naturally to the question which quantities should be related by formulas. A formula relates numerical values of different physical entities, and therefore these entities have to be measurable. The principle Galileo followed was to measure whatever was measurable, and to try to render measurable whatever was not yet so. His problem thus became to isolate those aspects of natural phenomena that are fundamental and susceptible to measurement.

Descartes had already singled out matter moving in space and time as the fundamental phenomenon in nature. Therefore, Galileo tried to isolate the characteristic qualities of matter in motion that can be measured and then related by mathematical laws. By analyzing and reasoning about natural phenomena, he decided to concentrate on such concepts as volume, time, weight, velocity, acceleration, inertia, force, and moment. In his choice of these particular quantities and concepts, Galileo once more displayed his genius, for the ones he chose are not at first the most important, nor is measuring them trivial. A concept like inertia is not even clearly inherent in matter, and makes itself known only indirectly; a concept like moment had to be created. Nevertheless, these concepts became extremely important in the elucidation of many of nature's secrets.

Although Galileo carried out experiments to a striking effect, we should not conclude that experiments began to be carried out on a large scale, or that the experimental method at once became a new and decisive driving force in science. It became so only in the 19th century. It was nonetheless revolutionary and altogether crucial for the evolution of science that Galileo acted as spokesman for the view that physical principles should be based on experience and experiment. Galileo himself was not in doubt that ultimately one could discover the true principles that God had employed in the formation of the universe. However, by opening the door to the experimental method, Galileo at the same time raised the question whether—if the fundamental principles of science are to be derived from experience—the axioms of mathematics should also be derived from experience? Galileo left this question unanswered.

To reach into the heart of natural phenomena, Galileo promoted and practiced yet another principle, namely, the principle of *idealization*. By this, he meant the disregarding of trivial or less significant factors. A ball falling to the ground thus encounters air resistance, but for a fall of a few hundred meters, air resistance is small and can for most purposes be discounted. In the same manner, a relatively compact object, even if it has extension and shape, can for most intents and purposes be regarded as a point of mass, that is, the mass is regarded as though it were entirely concentrated in a single point. He also disregarded secondary qualities of matter, such as taste, color, and odor, in contrast to dimension, form, mass, and motion. In other words, he adopted the philosophical doctrine that distinguishes between primary and secondary qualities of matter, as Descartes had also done. Form, mass, and motion are thus the primary, or the fundamental, physical qualities of matter.

Thus, what Galileo advocated was disregarding coincidental or less significant effects in order to get at the crucial relations. He started from observations, and so imagined what would happen if all resistance was removed, or in other words, if bodies fell in empty space (a vacuum). Here, he hit upon the principle that in a vacuum, all bodies fall in accordance with the same law. Since he had observed that the motion of a pendulum was only slightly affected by air resistance, he experimented with pendulums to confirm his principle. In the same way, since he also suspected that friction was a secondary effect, he experimented with smooth balls rolling down a smooth inclined plane to discover laws concerning frictionless motion. In this manner, Galileo carried out experiments and drew conclusions from them, disregarding facts that were relatively less significant in the interpretation of the experiments. In part, his greatness consisted of posing the right questions about nature.

By disregarding air resistance and friction, and imagining bodies in motion in empty space, Galileo was not only at odds with Aristotle, but also with Descartes. When Galileo sought the fundamental laws of motion in nature, he employed the method of idealizing or abstracting the essential characteristics of his objects, thus acting just as a mathematician does in studying forms from the real world. The mathematician disregards molecular structure, for example, and color and thickness of lines in order to derive certain fundamental characteristics, and then concentrates on these. In the same way, Galileo thrust to the core of the fundamental physical factors. The mathematical method of abstraction, initially a step away from the real world, leads paradoxically enough back to the real world to greater profit than if all the factors of a phenomenon were taken into account.

Galileo was also sage with respect to another matter. He did not attempt, like scientists and philosophers before him, to comprehend all natural phenomena, but selected a few fundamental phenomena and studied these intensively. Galileo displayed the restraint of a true master.

Galileo's program thus comprised four principal elements. The first was to seek quantitative descriptions of physical phenomena and to work these into mathematical formulas. The second was to isolate and measure the most fundamental quantities in the phenomena. It was these quantities that were to be the variables in the formulas. The third was to build up science deductively on the basis of fundamental physical principles. The fourth was to idealize.

To bring this program to fulfillment, Galileo had to discover fundamental laws. This search was an immense task, for he had to break once more with his predecessors. His entry into the study of matter in motion had to take account of an Earth that moved in space and rotated around its own axis, and these facts in themselves invalidated most of the only known significant mechanical system of the Renaissance, namely, Aristotle's mechanics. All Galileo's research was resolved ultimately in his three famous laws of falling bodies, of which we have cited earlier the formula for the distance of a fall. Galileo made other mathematical contributions, like his mathematical description of the motion of a projectile, but the core of his work is nonetheless his methodology. With his *Discorsi* of 1638, Galileo steered modern physical science onto its mathematical track, laid the foundations of modern mechanics, and offered the pattern for all natural scientific thought to come.

5. Newton and Gravitation

It was Newton who adopted Galileo's methodology and gave unsurpassed demonstrations of its efficacy. The very year Galileo died, Isaac Newton (1642–1727) was born on a small farm in the English village of Woolstorp. Apart from a strong interest in mechanical contrivances, Newton displayed no particular talent as a youth. On the purely negative grounds that he showed no interest in farming, his mother sent him to Cambridge, where he matriculated at Trinity College in 1661. Although he had access there to the works of Descartes, Copernicus, Kepler, and Galileo, and could attend lectures of the renowned mathematician Isaac Barrow, Newton does not seem to have gotten much benefit from it. He was even judged to be weak in geometry, and there came a point in time when he nearly switched over from the study of science to law. Four years of introductory studies at Cambridge ended rather unimpressively.

At this point, a plague broke out in the London area, and the university was closed. Newton therefore spent the years 1665 and 1666 at the family home in Woolstorp. It was during this period that he began his great work on mechanics, mathematics, and optics. He saw that the law of gravity was the key to an all-embracing mechanics; he derived a general method of handling problems from mathematical analysis; and through experimentation, he made the epoch-making discovery that white light, such as sunlight, is actually composed of all the colors, from violet to red. Newton himself said that during these two plague years he was at the height of his creative powers, and more mathematically and philosophically inclined than at any time thereafter.

In 1667, Newton returned to Cambridge and became a Fellow of Trinity College. In 1669, Isaac Barrow retired from his professorship in mathematics to devote himself to theology, and Newton was named Barrow's successor. It seems that Newton had no great success as a teacher, for few students attended his lectures, nor did anyone note the originality of the material he presented. In 1684, his friend, the astronomer Edmond Halley (known for Halley's comet), urged him to publish his work on gravitation, and assisted him both editorially and financially. In this way, the classic of natural science, *Philosophiae Naturalis Principia Mathematica* (*The Mathematical Principles of Natural Philosophy*), often referred to simply as *Principia*, came to be published. Following its publication, Newton became famous. *Principia* was brought out in three editions, and popular versions were soon generally available. It was actually necessary to popularize *Principia*, for the book is extremely difficult to read. The greatest mathematicians worked for more than a century to elucidate the material of the book fully.

Newton fully acknowledged his predecessors; he did not feel there was anything special about his work. In his old age, he is said to have told his nephew, "I do not know how I may appear to the world, but to myself I seem merely to have been a little boy, playing on the beach, and diverting myself now and then in finding a smoother pebble or an especially pretty shell, while the great ocean of truth lay undiscovered before me." This picture of Newton is the classic version of him as a modest, affable man. In *A Brief History of Time*, however, published in 1988, his successor in the professorship at Cambridge, the famous astrophysicist Stephen W. Hawking, sketches a somewhat different portrait of Newton: He may in truth have been not so affable.

Of the great contributions Newton made in his youth, it is his natural philosophy and his work on gravitation that primarily will occupy us here. Newton's natural philosophy is more explicit than the scientific program outlined by Galileo: From clearly verifiable phenomena, laws of nature are to be extracted, describing nature in the precise language of mathematics. By employing mathematical arguments to these laws, new ones may be derived. Like Galileo, Newton too wished to know how God Almighty had created the universe, but he did not entertain the hope of coming to understand the mechanism behind all phenomena. For Newton, as for Galileo, the mathematical principles were quantitative principles. As Newton says in *Principia*, it is his object to discover and reveal the precise way in which "all things had been ordered in measure, number, and weight."

In his effort to describe nature, Newton's most famous contribution was to combine heaven and Earth. Galileo had explored the heavens in a way no one before him had been in a position to do, but his success in describing nature mathematically was limited to motion on, or close by, the surface of the Earth, even if his contemporary Johannes Kepler had in his three famous mathematical laws of motion of the heavenly bodies succeeded in greatly simplifying the heliocentric theory. The two branches of natural science, the terrestrial and the celestial, were apparently independent of each other; it was a challenge to all great scientists to discover a connection between the two. This extaordinarily significant connection was found by Newton.

There was good reason to believe there might be an underlying unifying principle, for just as a weight swung around at the end of a string does not fly off along a straight line because the hand exerts a force pulling it inwards, there must be a force that continuously deflects the planets from moving in a straight line. It was apparently the sun itself that exerted an attractive force on the planets. Scientists of Newton's time recognized also that the Earth attracts bodies. Thus, since both the Earth and the sun attract bodies, the possibility of unifying the two effects within the same theory had already been proposed and discussed at the time of Descartes. Newton reformulated this idea into a mathematical problem, and without determining the physical nature of the forces involved, he solved this problem by means of brilliant mathematics. The story goes that Newton got the idea of referring the Earth's pull on bodies and the sun's pull on the Earth to the same schema through an apple's fall from a tree. This is supposed to be authentic, even though the great mathematician Karl Friedrich Gauss (1777–1855) said that Newton only told that story to brush off the untutored persons who pestered him about how he discovered the law of gravity.

Newton began by considering the problem of propelling a projectile horizontally from the top of a mountain. Galileo had already worked out in detail exactly this sort of problem and had proved that the resulting trajectory of the projectile is a parabola. If the horizontal velocity of the projectile is increased, the trajectory always remains a parabola, although the projectile flies further out. Galileo had considered projectiles shot only a short distance, and had therefore disregarded the curvature of the Earth. Newton's first thought therefore was that if the projectile were shot horizontally from the mountain top with greater and greater velocity, it would land as before, further and further away, but if it were shot with sufficient velocity, it would make a complete orbit of the Earth and perhaps continue to orbit the Earth forever. See Figure 4.6. In *Principia*, Newton argues that, by analogy to the experiment

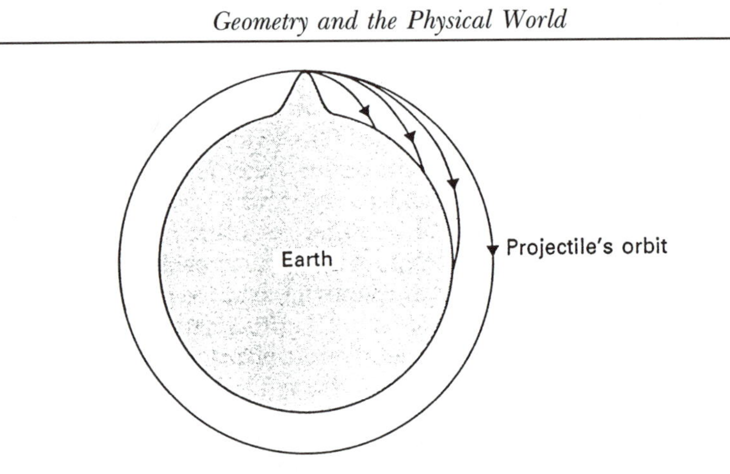

Figure 4.6. A projectile is shot out horizontally from the top of an "extraordinarily high" mountain on the Earth. At a suitably high horizontal speed, the projectile goes into orbit around the Earth.

with the projectile, it must be gravity, or some other force, that holds the moon in its orbit around the Earth; and he continues with the claim that if the Earth with its gravitation can hold the moon in its orbit, the sun with its gravitation must also be able to hold the planets in their orbits around it. Newton had thereby reached the conclusion that the same force that pulls at objects close to the surface of the Earth also holds the moon in its orbit around the Earth and the planets in their orbits around the sun. He had thus boldly integrated terrestrial and celestial mechanics.

This argument for the connection between terrestrial and celestial phenomena is purely qualitative. Newton gave a more quantitative argument as well, however, that in simple terms runs as follows. The moon's orbit around the Earth is more or less a circle. Since the moon, designated M in Figure 4.7, does

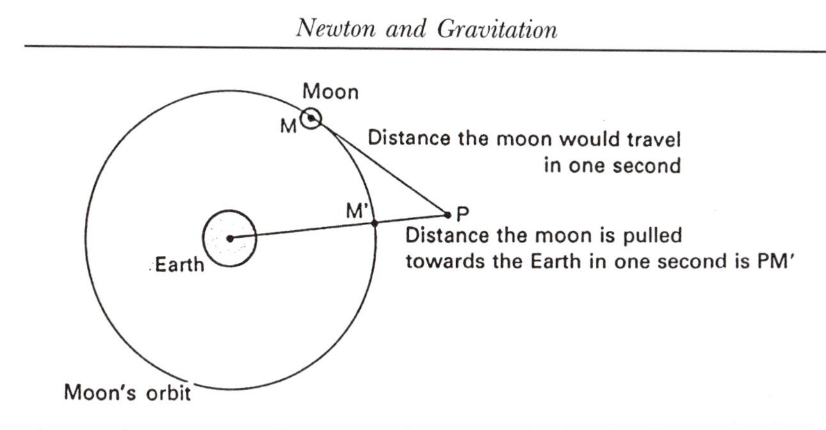

Figure 4.7.

not pursue a straight line such as MP, it apparently is drawn towards the Earth by some force. If MP is the distance the moon will have traveled in 1 second in the absence of any affecting force, then PM′ is the distance the moon is pulled towards the Earth in this second. Newton now used PM′ as a measure for the Earth's attractive force on the moon. The corresponding length in the case of a body close to the Earth's surface is 4.9 meters, since that is the distance a body falls in the first second after it is dropped from a tower. Newton wished to prove that the same force could account for both the moon's fall PM′ towards the Earth, and the 4.9 meters a body falls when close to the surface of the Earth.

Rough calculations had led Newton to believe that the force that attracts bodies to each other depends on the square of the distance between the centers of the two bodies, and that the force thus diminishes as the distance increases. The distance between the moon's center and the Earth's center is about 60 times the Earth's radius. Therefore, the Earth's effect on a body close to the surface of the Earth ought to be about

$60 \cdot 60 = 3600$ times greater than the Earth's effect on the moon—in other words, the moon ought to be pulled towards the Earth at about $(4.9)/3600 = 0.00136$ meters per second. By carrying out some trigonometric calculations, Newton found that the moon in fact is pulled towards the Earth at something very close to this result. He achieved thereby a persuasive confirmation that all bodies in the universe attract each other in accordance with the same law.

More meticulous investigations led Newton to the precise formula for the attractive force between two arbitrary bodies, namely, the famous *law of gravity*:

$$F = \frac{k \cdot M \cdot m}{r^2},$$

where F is the attractive force, M and m are the masses of the two bodies, r is the distance between them, and k is a universal constant that is the same for all bodies. For example, let M be the Earth's mass and m the mass of an object near the Earth's surface. In this case, r is the distance from the Earth's center to the object.

Summing up his entire work on terrestrial and celestial motion, Newton formulated in *Principia* what are now known as *Newton's three laws of motion*—even if the first two laws actually had already been formulated by Descartes and Galileo. The three laws are formulated as follows:

(1) A body remains at rest or has constant velocity (speed and direction) if it is not acted upon by any external force.

(2) The total force on a body is the product of its mass and its acceleration.

(3) When two bodies act upon each other, the force on the first body from the second body is equal to and opposite the force on the second body from the first body (the law of action and reaction).

After Newton had obtained some confirmation of the law of gravity from his calculations of the moon's motion, he subsequently demonstrated that it also could be used for motions on or close to the surface of the Earth. By exploiting the law of gravity and the law stating that the force acting on a body is proportional to the body's acceleration (law 2), he could deduce mathematically that close to the Earth's surface, the acceleration of a body falling only under the influence of gravity is the same for all bodies. This was, of course, nicely consistent with the law of falling bodies that Galileo had derived experimentally. It can easily be ascertained that the velocity of a falling body increases every second by 9.8 meters per second.

With this achievement, demonstrating that the laws of motion and the law of gravity are the fundamental laws, Newton had realized one of the major goals of Galileo's program; but he accomplished another striking feat as well in demonstrating, through a series of far-reaching mathematical deductions paradigmatic of the strength of mathematics, that Kepler's three laws of planetary motion follow from the first two laws of motion and the law of gravity.

The works of Galileo, Newton, and their successors were in overwhelming measure mathematical in the sense that they exploited mathematics to gain knowledge of physical phenomena that could not have been acquired through the world of experience. Of course, it was often observations from the physical world, like those of falling bodies or the motions of the heavenly bodies, that suggested the mathematical problems, and in many cases the results gleaned were susceptible to corroboration by experiment or observation. A striking example is the purely theoretical prediction of the planet Neptune's existence. Galileo had seen it already in 1613, but thought it was a star. Inexplicable deflections in the motions of the planet Uranus, observed around 1820, were ascribed to the gravitational pull on Uranus from an unknown planet, dubbed Neptune. The precise orbit of the planet and its mass were calculated by the English astronomer Adams in 1841. The calculations were initially ignored to some extent, but in 1846 they were sent to the German astronomer Galle, and on the evening of the very day he received the calculations, he observed Neptune. Without the aid of mathematics, Neptune would hardly have been discovered in 1846, with the telescopes available at the time.

Despite these triumphs, all attempts to understand the physical nature of gravity have gone awry. Galileo had already sought a physical explanation, and Newton too harbored the ambition to find one. We can state, however, that even if we now can send vehicles into space on the basis of Newtonian mechanics, and even if the mathematical theory works perfectly, the true physical nature of gravity remains a mystery nevertheless.

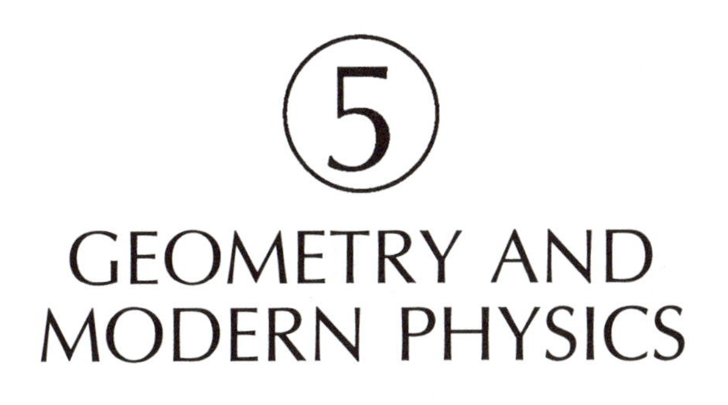

5 GEOMETRY AND MODERN PHYSICS

In the 17th and 18th centuries mathematicians and physicists developed a series of outstanding mathematical theories that extended human knowledge of many phenomena in nature. Newton's law of gravity especially was a major breakthrough. Theories of heat conduction, hydrodynamics, and elasticity were also developed. It is true of all these achievements that they derived from phenomena perceived by the senses, very much in the spirit of Aristotle. Of course, the mathematical theories went beyond the observations, and they led, as we saw in Chapter 4 in the case of the laws of motion, to new concepts apparently not rooted in physical reality. Nevertheless, predictions on the basis of these theories corresponded exceptionally well to experimental results.

Although it was believed that nature was organized mechanically, no physical explanation had successfully been given for how gravity and light operated. In the case of light, scientists assumed it was propagated through an *ether*, a concept that should permit a mechanical explanation, even if some details were still missing. So far as gravity was concerned, the nature of its operation was still ultimately a mystery. Newton's theory, however, and the later contributions to it by, *inter alia*

Euler, d'Alembert, Lagrange, and Laplace, were so remarkably precise in their mathematical descriptions and predictions of a whole series of astronomical phenomena that scientists were ecstatic.

Developments in the 19th and 20th centuries, to which we now turn our attention, raised fundamental questions concerning the nature and content of our physical world. The first of these developments dealt with electricity and magnetism, and it added yet another phenomenon to our physical universe, namely, the electromagnetic theory, to which we shall give a brief introduction in the first section of this chapter.

Physical space provides us not only with the fundamental concepts of Euclidian geometry, but also with a framework in which to imagine the much more general types of spaces that continually appear in mathematics, and that we have already encountered some of in Chapters 2 and 3; but the connection between the development of mathematical and physical concepts goes much further, for it was while working on the formulation of his laws of classical mechanics that Newton was led to develop fundamental concepts of mathematical analysis, such as continuity and differentiability. Indeed, even very fundamental mathematical concepts like that of the real numbers have their origin in measurements of separation in time and space.

All this is carried to its logical conclusion in Albert Einstein's theory of relativity from the beginning of this century, where both length and time are geometrical quantities whose measurements are closely tied to space. In studying the theory of relativity, it is shocking, therefore, to discover that when all has been said and done, Euclidian geometry does not describe physical space in the most precise way. From Euclidian geometry, in fact, has sprung a much more flexible geometry known as *differential geometry*, in terms of which Einstein's theory of relativity is formulated. When the general theory of relativity was presented in 1915, it was a very daring theory, but it has proven by now to agree exceptionally well with observations. Thus, if we wish to understand how the universe is structured, we cannot avoid the theory of relativity, which is to be the subject of Sections 2, 3, and 4.

In stars and galaxies, in atoms and molecules, and in all living organisms, there are but four forces that carry out all of nature's work. These four forces cause all known interactions, collisions, commotions, and other reactions at any moment and at any location. On an average day, one encounters all the forces, or at least their effects. *Gravity* is the first of the forces, the force that holds us to the Earth. The second force is the *electromagnetic force*, to whose existence electromagnetic phenomena testify. It holds organisms, and everything else built of atoms, tightly together. The force is transmitted by electromagnetic waves, and among its forms are light rays, X-rays, and radio waves. The third and fourth forces of which we know are nuclear forces with ranges much smaller than in atoms. Nevertheless, they often make themselves known in the macroscopic world through effects that reveal them to the naked eye. These forces are called, respectively, *strong* and *weak interactions*. Work by the strong interactions gives nuclear energy, either in the form of fusion or of fission, and it delivers the driving force that lights up the sun and the stars. The weak interactions are more subtle and difficult to detect, but they are responsible for the decay of certain particles, and deliver, for example, the driving force to the glimmering green luminescence visible in the dark on a watch with a radium dial.

Ever since Einstein, it has been the goal of physicists to create a unified theory encompassing the four known natural forces. There is now reason to hope that the so-called gauge theories, or more generally the string theories, developed since the mid-1950s, can yield such a theory; this is the subject of Sections 5, 6, and 7. In the formulation of these theories, concepts of differential geometry also play a crucial role.

Modern physics utilizes the very latest and most advanced branches of mathematics, so fair notice is hereby given of the even greater demands this chapter will place on the reader's mathematical imagination and patience.

1. Maxwell and the Electromagnetic Theory

The connection between electricity and magnetism was discovered in 1820 by the Danish physicist Hans Christian Ørsted (1777–1851). The story goes that one day in his study, Ørsted discovered that a magnetic needle was deflected when a current was passed through a wire in its vicinity. He rushed in enthusiastically to tell his assistant of his discovery. Thoroughly unimpressed, however, the assistant restricted himself to remarking dryly that the needle did this every time. When Ørsted presented his discovery to the Royal Danish Academy of Sciences and Letters, he was asked whether he believed this discovery would have any practical significance. Ørsted is supposed to have replied that he rather doubted it, but that it was an interesting phenomenon nonetheless.

Further connections between electricity and magnetism were discovered by the French physicist André-Marie Ampère (1775–1836), who found in 1821 that two parallel wires through which an electric current is sent behave like two magnets: If the currents travel in the same direction, the wires

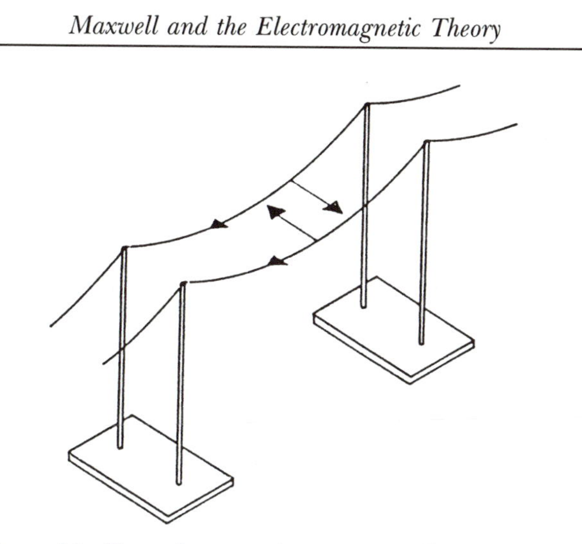

Figure 5.1. Two wires carrying current affect each other.

attract each other, and if they travel in opposite directions, they repel each other. See Fig. 5.1. In 1831, the English bookbinder Michael Faraday (1791–1867) and the American schoolmaster Joseph Henry (1797–1878) independently discovered the opposite effect, namely, that an alternating magnetic field induces current in a coil. Faraday's arrangement is shown in Fig. 5.2. This phenomenon is now called *electromagnetic induction.*

The great figure in electromagnetism is the mathematical physicist James Clerk Maxwell (1831–1879). He invented the color photograph, and contributed to the formulation of the kinetic theory of gases. His major achievement, however, consists of his crystallization of all the known electrical and magnetic phenomena into a single theory. It is largely overlooked in the international physical literature that the Danish physicist

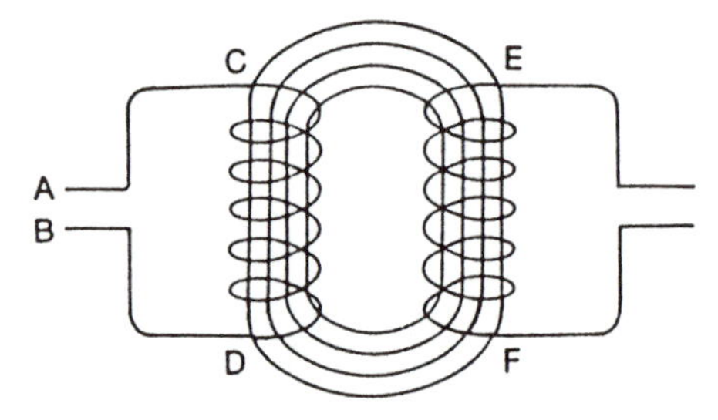

Figure 5.2. An alternating current between poles A and B produces an alternating magnetic field around the coil CD (indciated by the oval lines). This alternating magnetic field induces an alternating current in coil EF.

L. V. Lorenz (1829–1891), independently of Maxwell, also presented a unified theory of electrical and magnetic phenomena.

At the beginning of the 1850s, Maxwell became strongly influenced by the work of William Thomson, who was later ennobled Lord Kelvin (1824–1907). Among other things, Thomson had laid the foundations of the mathematical theory of propagation of waves in fluids, from which Maxwell profited. Thomson was also an adherent of mechanical explanations of electrical and magnetic phenomena, and he experimented with analogies to these subjects from the theories of fluid flow, heat flow, and elasticity. He employed analogies from these areas to the ether, which was then an unspecified—but generally accepted—medium for the propagation of light and heat, and to which we shall return. Thomson regarded the ether as a *field* of forces that can act on test particles in each point of space. The concept of a field had previously been proposed by the mathematicians Cauchy, Poisson, and Navier. The field is, so to speak, there all the time, but makes itself known only when a small electric charge or a small magnet is

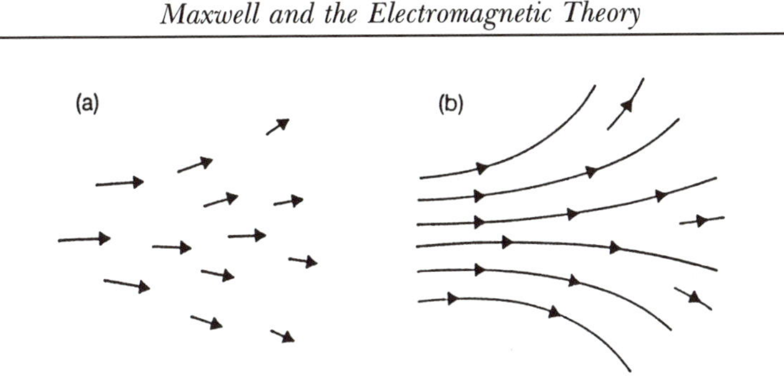

Figure 5.3. A vector field can be represented by drawing a set of arrows as in (a), where the length and direction of each arrow determine the vector field at that point from which the arrow originates. A vector field can also be represented as in (b), by drawing a system of curves (field lines) that are tangential to the direction of the vector field, and where the density of the field lines are proportional to the magnitude of the vector field.

introduced into it. Regarded mathematically, a force field is nothing but a vector field, that is, a vector introduced at each point in space. Many physicists are fond of drawing the so-called *field lines*, which are tangential curves to the field. The density of the field lines are taken as a measure of the field's strength. See Fig. 5.3. Maxwell sought a mechanical explanation of the action of the ether, but neither he nor Thomson was successful on this score. Nevertheless, Thomson introduced what is now called the *field concept*, and Maxwell adopted this concept.

A field is of an entirely different character from the "well-known" physical forces, like elastic forces, pressure forces, friction forces, etc., and can best be understood in reality as a purely mathematical concept. It is nonetheless possible to illustrate the magnetic field with the aid of iron filings. Most people have tried placing a magnet on a table beneath a sheet of paper, and spreading iron filings on the paper. If the paper is tapped lightly, the filings, which under the influence of the magnetic field have themselves become tiny magnets, organize themselves along the field lines in a quite striking pattern. In the same way, we can demonstrate the magnetic field around a current-bearing electric wire. If a copper wire is pulled through a hole in the paper, and a current is passed through the wire, the filings will organize themselves in circles around the wire. If the wire is wound into a coil by passing it several times through the paper, a magnetic field is formed whose field lines pass through the entire coil, and from there in large arcs around its outside and back to their starting point.

In 1865, Maxwell published his major work, *A Dynamical Theory of the Electromagnetic Field*, in which he disregarded all the mechanical models and presented the relevant mathematics in several equations that related the electrical field **E** and the magnetic field **B**. The mathematical formulation convinced him that such fields could propagate over large distances.

We can only briefly review Maxwell's equations, which are normally expressed as four laws. They are concisely formulated with the aid of mathematical vector analysis, and have a rich geometrical content. The mathematics in Maxwell's model of electromagnetism is not elementary, but a total comprehension of the mathematics of the equations is fortunately not requisite to understanding the following.

First, we give the laws for a vacuum. To present the essentials clearly, we have left out the constant factors that normally appear in the laws:

1. $\operatorname{div} \mathbf{E} = \rho$ (Gauss's law).

2. $\mathbf{curl}\, \mathbf{B} = \mathbf{j} + \dfrac{\partial \mathbf{E}}{\partial t}$ (Ampère–Maxwell law).

3. **curl E** $= -\dfrac{\partial \mathbf{B}}{\partial t}$ (Faraday's law).
4. div $\mathbf{B} = 0$ (Absence of magnetic monopoles)

In the preceding formulas, **E**, **B**, and **j** denote, respectively, the electrical field, the magnetic field, and the electrical current density. The operations div (divergence) and **curl** are operations from vector analysis, which at every point of a field measure, respectively, the flux of the field per unit of volume out through a small sphere around the point, and the rotation vector for a spatial vortex, corresponding to the field, in a neighborhood of the point. The quantities

$$\frac{\partial \mathbf{E}}{\partial t} \quad \text{and} \quad \frac{\partial \mathbf{B}}{\partial t}$$

measure the variation in time of **E** and **B** respectively. Finally, ρ denotes the density of the electrical charge.

Maxwell's equations are so-called partial differential equations, whose theory is an important part of mathematical analysis. The electrical and magnetic fields **E** and **B** are functions both of the spatial coordinates and of the time t.

A qualitative verbal formulation of the four laws is:

1. The flux of the electrical field out from a finite volume is proportional to the electrical charge within the volume.
2. An electrical current or an alternating electrical flux gives rise to a magnetic eddy.
3. An alternating magnetic flux gives rise to an electrical eddy.
4. The total magnetic flux out from a finite volume is always zero.

The electrical flux and the magnetic flux from a volume enclosed by a surface measures the number of field lines of respectively the electrical field and the magnetic field out through the surface.

It follows from Maxwell's theory that if an electrical current is allowed to fluctuate back and forth in a conductor, the electromagnetic fields, which of course alternate in time with the current, will "tear themselves free," so to speak, from the conductor and propagate in space as *electromagnetic waves*. From a mathematical point of view, the electromagnetic waves are simply the solutions to Maxwell's equations. The *frequency* of the electromagnetic waves (the number of complete alternations per second) is the same as the frequency of the alternating field, and Maxwell was able to prove mathematically that in a vacuum they propagate at a velocity of about 300,000 km. per second. Maxwell's first and greatest theoretical discovery is that electromagnetic waves can travel across distances of thousands of kilometers from a source to a suitable receiver.

In the course of his mathematical work, Maxwell thereby made a sensational discovery on the nature of light. Light had been studied from the time of the Greeks, and following innummerable experiments, there were now two competing physical theories. One theory (the *corpuscle theory*), proposed by Newton, around 1650, assumed that light consists of small invisible particles (corpuscles) that are radiated by all luminous bodies and move along straight lines. The other theory (the *undulation theory*), proposed by Huygens, also around 1650, described light as a wave motion in a substance (the ether) filling the universe. Both theories explained reflection and refraction of light (i.e., the change in direction that occurs when, for example, light passes from air into water) somewhat satisfactorily. Diffraction (scattering) of light, however, such as occurs when light moves around an obstacle like a disk, is explained more reasonably by a wave theory analogous to the theory that accounts for the bending of water waves around the jetties of a harbor entrance. At the beginning of the 19th century, Thomas Young (1773–1829) and Augustin Fresnel (1788–1827) argued strongly for the wave theory in a medium (the ether), which they specified no further.

Another early development in the history of light should be mentioned. In 1673, the Danish astronomer Ole Rømer (1644–1710) proved that the speed of light is finite, and he actually found a close numerical approximation of its value. He discovered that eclipses of Jupiter's moon Io were about 16 minutes behind schedule when observed while the Earth was farthest away from Jupiter, compared with observations made while the Earth was closest to Jupiter. The difference in the distance traveled by the light from Io in the two situations is approximately equal to the diameter of the Earth's orbit, roughly 300 million km, which gives a speed of light of about 312,500 km per second. More precise measurements in the 19th century proved that the speed of light is nearly 300,000 km per second. (The meter is now defined such that the speed of light in a vacuum is fixed at 299,792,458 meters per second.)

In his mathematical investigations, Maxwell found that the speed of the electromagnetic waves was nearly equal to the speed of light, and since both electromagnetic radiation and light were wave motions, Maxwell was led to the conviction that light is an electromagnetic phenomenon. In 1862, Maxwell said, "We can scarcely avoid the inference that light consists in the transverse undulations of the same medium (namely, the ether) that is the cause of electrical and magnetic phenomena." He wrote a paper concerning this in 1868, and his theory, known as *Maxwell's electromagnetic theory of light*, has remained the standing theory of light ever since.

In Maxwell's day, physicists were partly aware of the existence and the characteristics of ultraviolet rays, which, though invisible to the eye, betray their presence by blackening photographic film. In addition, they knew about infrared rays, which are also invisible, but transmit heat, which is easily registered on a thermometer. Both of these types of rays occur in the sun's radiation. They can also be produced by sending electrical current through special filaments (thin wires) in the same way visible light can be produced by passing current through tungsten. The conjecture that infrared rays and ultraviolet light are electromagnetic waves was easily confirmed experimentally, and it was discovered that infrared rays have somewhat lower frequencies than visible light, while ultraviolet light has somewhat higher frequencies.

Soon, more and more pieces of the electromagnetic puzzle fell into place. In 1895, a German physicist, Wilhelm Konrad Röntgen (1845–1923), discovered rays he called X-rays, and these soon were identified as electromagnetic waves with frequencies higher than the frequencies of ultraviolet light. Finally came the discovery of gamma rays, which stem from radioactive substances, among other things, and it was found that gamma rays were also electromagnetic waves, with frequencies even higher than those of X-rays. The electromagnetic spectrum comprises wavelengths from 10^{-14} to 10^{8} meters, that is, a spectrum of more than 10^{22}, an unimaginably large number. (For sound waves, an octave corresponds to a doubling of the frequency. Since 10^{22} is approximately the same as 2^{73}, we can, by analogy with sound waves, say that the electromagnetic spectrum spans more than 72 octaves. The visible band covers only one of these octaves, so our vision is very limited; but we have other instruments that can detect infrared rays, ultraviolet light, X-rays, and gamma rays.)

In 1873, the year Maxwell published his *magnum opus* on electromagnetism, virtually all physicists were skeptical about electromagnetic waves. At the very least, they found the concept difficult to imagine. An exception was the Dutch physicist Hendrik Antoon Lorentz (1853–1928), who tried unsuccessfully to produce the different types of waves experimentally. He did show in his doctoral thesis of 1875, however, that Maxwell's theory explains reflection and refraction of light better than any other theory then in existence.

There clearly was a need for experimental confirmation, for Maxwell's laws involve some physical assumptions, and mathematical predictions on the basis of such assumptions might not lead after all to correct physical results, since the assumptions themselves might be incorrect. In 1887, about 25 years after Maxwell had predicted the existence of electromagnetic waves, the German physicist Heinrich Hertz (1857–1894) produced electromagnetic waves and received them in a circuit placed some distance from the source. For a long time, these waves were called Hertzian waves—nothing other than the radio waves used today in thousands of ways. This was a striking corroboration, and it was soon followed by other applications.

When Maxwell proved that the electromagnetic waves travel at the speed of light, he concluded that these waves must propagate through the ether, for since Newton's time the ether had been accepted as a medium for the transmission of light. Since the waves travel at an enormous velocity, the ether must be extremely rigid, for the more rigid a body is, after all, the faster waves can travel in it. On the other hand, the ether—if it extends throughout the universe—must be completely transparent, and the planets must travel through it without friction. The conditions thus placed on the ether are filled with contradictions. In addition, the ether can neither be felt, smelled, nor isolated from any other substance; but even though such a medium is physically unacceptable, the ether theory was nevertheless finally discarded only by Einstein in connection with the theory of relativity.

Indeed, one can say that we have no way of *mechanically* accounting for the action of the electrical and magnetic fields; that is, we cannot "explain" these fields by other physical concepts. Nor do we have any immediate perception of the electromagnetic waves as waves. Only when, for example, we introduce receiving antennae into the electromagnetic fields do these make themselves known. Nevertheless, we send radio waves containing complicated messages across thousands of miles. The only thing we do not know is what exactly is propagated in space. It is equally thought-provoking that these waves are all around us; we need only flick on the radio, after all, to receive waves sent out from near and far. Normally, our senses do not have the slightest awareness of their presence. The lack of understanding of the physical nature of electromagnetic waves was dissatisfying for many of the major figures behind the theory. Thus, Lord Kelvin, in a speech in 1884, said that he was not satisfied with Maxwell's work, and that he could never be satisfied until he could construct a mechanical model of a thing. What was missing was a mechanical model for the ether.

The precise and comprehensive presentation of electromagnetism is a mathematical formulation, and no one valued the thoroughly mathematical character of the electromagnetic theory more highly than Maxwell himself. Although he tried almost desperately to develop a mechanical theory for the electromagnetic phenomena, he left most of this material out of his classic work in 1873, *A Treatise on Electricity and Magnetism*, and emphasized here the finely polished mathematical theory. His attempt at diluting the mathematics of electromagnetic fields with intuitive, more easily understandable explanations was not very successful, and one might well say that while in some branches of physics it is possible to fit the mathematical theory to physical facts, about the best one can do with the electromagnetic theory is to fit inadequate physical pictures to mathematical facts.

Maxwell set the standards for the practice of modern mathematical physics, and they are indeed principally mathematical. Maxwell's electromagnetic theory exceeds even Newton's theory of gravitation in that it includes a large group of highly dissimilar phenomena in a single mathematical theory.

The electromagnetic theory provides another illustration of the power of mathematics to reveal nature's secrets; but our inability to explain the electromagnetic phenomena qualitatively or materially stands in sharp contrast to the precise quantitative description we have obtained through Maxwell's theory. In the same way Newton's laws of motion offered scientists a tool with which to work with matter and forces without explaining any of these things, Maxwell's laws put scientists in a position to accomplish great things with electrical phenomena despite a highly deficient comprehension of their physical nature.

Mathematical theories form the basis of our understanding of both gravitation and electromagnetism. The underlying physical causes remain unknown, and we have to be satisfied with representing the phenomena mathematically. This can seem unsatisfactory, of course, but as the famous philosopher and mathematician Alfred North Whitehead (1861–1947) has said, "The paradox is now fully established that the utmost abstractions (of mathematics) are the true weapons with which we control our thinking about concrete facts." In 1931, Einstein described the change that had occurred in the conception of the physical world after Maxwell as, "the most profound and the most fruitful that physics has experienced since Newton."

The mathematical physicists at the turn of the century were proud of the results they had achieved, and were in general rather satisfied with the physical theories, but despite the remarkable achievements, many of them understood nonetheless that great problems remained to be solved.

2. Einstein's Theory of Relativity

The conception of light as wave motion seemed to require a medium, an ether, in which this wave motion could take place. Fresnel thought it was a question of mechanical vibrations in an elastic ether that was supposed to exist everywhere throughout empty space. When it was later found that in that case it had to be pure transversal vibrations, perpendicular to the direction of the propagation of light, Fresnel was eventually compelled to ascribe very peculiar characteristics to his ether to explain the propagation of light.

Fresnel and others came to the conclusion that the ether was at rest, and that, for example, the Earth in its orbital motion around the sun did not carry the ether along with it. One had to assume then that an "ethereal wind" passed continuously across the Earth, and that this ethereal wind could in one way or another be detected. Provided that it was possible to determine the velocity of the "ethereal wind", one would also be able to establish the Earth's absolute speed through the resting ether. Various attempts were made to obtain an answer to this question.

The American physicist Albert A. Michelson (1852–1931) devised an experiment along the following lines. If the theory of an ether at rest in which the Earth moves is correct, the speed of light measured relative to the Earth must be a little different in the direction of the Earth's motion from that in a direction perpendicular to it. In 1881 and 1887, he set up his experiment. Michelson's experimental apparatus is shown in Fig. 5.4. From a light source, a light beam is sent out a fixed distance in the direction of the Earth's motion, and reflected back again by a mirror. Another light beam is sent out simultaneously from the same light source an equal distance, but perpendicular to the first, and reflected back similarly by a mirror. When the light beams return, one beam ought to be delayed relative to the other because of the difference in their speeds, so that interference should occur when they merge again. Michelson's experiment proved clearly, however, that the speed of light was the same in both directions; no interference could be detected, nor could any difference be measured at the time at which the two beams returned, despite the fact that the time difference predicted by the theory was 100 times greater than the accuracy of measurement.

To explain this experimental result, Lorentz—and, independently, also Fitzgerald—proposed the bold theory that all bodies that move through the ether shrink a small amount in the direction of their motion. Purely mathematically, Lorentz was able to prove that if one let the contraction increase suitably with the speed, the resulting difference in the wavelength of the two light beams in Michelson's experiment will also increase, approaching the difference in speed dictated by the

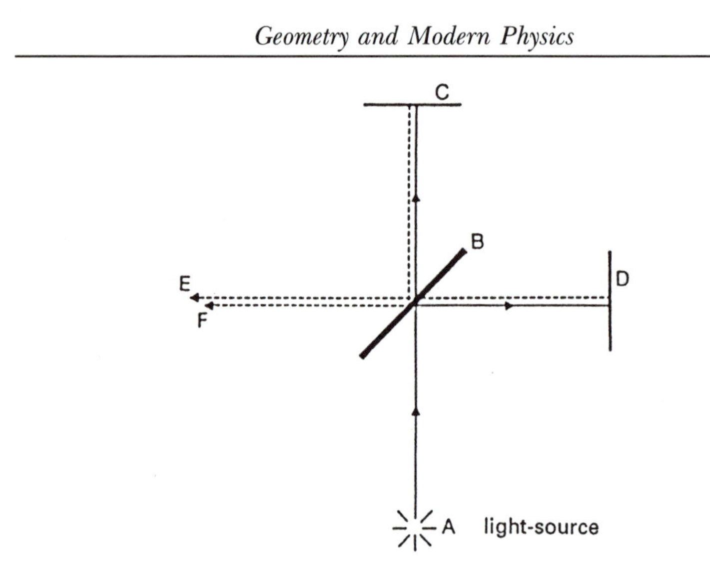

Figure 5.4. Sketch of Michelson's experimental apparatus. B is a specially silvered glass plate that splits a light beam coming from light source A into two beams perpendicular to one another. These beams are reflected by mirrors C and D, and at their return to B, they merge in beams E and F. The distances BC and BD are equal. Michelson observed no interference between the beams in E and F.

ether theory. The fact that no one had ever proposed a contraction like that required by Lorentz's theory was not remarkable, since all objects—yardsticks as well—are equally shrunk, and since furthermore the contraction at a velocity like the Earth's was only about one 200 millionth part.

Lorentz's theory was met with great skepticism, and it was naturally regarded as a far-fetched explanation. Posterity, however, has partly vindicated Lorentz. New experiments to detect the ethereal wind were conceived, but all yielded negative results. The unsuccessful attempts to determine the Earth's absolute velocity via the ether was finally explained by the special theory of relativity proposed by the German-born physicist Albert Einstein (1879–1955) in 1905. To understand Einstein's ideas, it will be necessary to discuss briefly the shifting opinion regarding absolute and relative motion through the years.

The English astronomer Sir Arthur Eddington has illustrated the concepts of absolute and relative motion in the following manner. Imagine a man in a completely closed elevator that is falling towards the Earth from a great height at the acceleration that gravity would give it. The man himself is ignorant of the location of the Earth, and takes no notice of the elevator's increasing speed. Now he puts his hand in his pocket and takes out an apple, extends his hand, and lets go of the apple. Under normal circumstances, the apple would fall directly to the bottom of the elevator, but in this case it does not happen. The apple remains suspended in the air for the simple reason that it is already falling just as fast as the whole elevator, and cannot fall any faster. For the man in the elevator, therefore, it will look as if the apple is not in motion at all. Now he takes another apple from his pocket, extends his hand again and places it in the space inside the elevator, this time a little closer to himself, but at the same height as the first apple. At first, it seems as if nothing is happening. The apples hang neatly suspended in space; but as time passes, the man will notice that the two apples have approached each other, as though some attractive force existed between them.

To an observer watching all this at a distance, it looks quite different. He sees that elevator, man, and apples alike are falling through space, drawn by the Earth's gravity. When the apples approach each other, it is because they are both attracted towards the center of the Earth—the distance between them decreases steadily. To the man in the elevator, however, it seems as if the two apples attract each other. Since the man in the elevator believes he is at rest (he is ignorant of the elevator's fall), he invents, so to speak, the attractive force between the two apples, but had he known he was falling in an accelerated motion, he would probably have seen at once how to explain the apples' motion towards each other. What then are absolute motion and rest? The man in the elevator believes he is at rest, but were he in a position to look out he would see "the observer" in motion, while the observer would assert that he himself was certainly at rest, and that it was the man in the elevator who was falling.

Aristotle considered the Earth as being at rest, and therefore believed that the motion of bodies relative to the Earth was their absolute motion. Copernicus came to the conclusion that all motion is relative, while Newton believed there was an absolute motion. With Fresnel's ether theory at the beginning of the 19th century, the question of an absolute motion again became important, this time with respect to the resting ether, and with this as his foundation, Lorentz constructed his theories. He pointed out that if one assumes that the mass of a body as well as the forces between various bodies change with the velocity of the bodies through the ether just as their dimensions change (contraction), then one obtains the result that natural phenomena seem to occur according to exactly the same laws in a system in motion and in a system at rest, provided of course that measurements are made with the system's own yardsticks. An observer at rest will, with his instruments, be able to confirm contractions, changes in mass, etc. in a system in motion. In the Lorentzian theory, a body's mass is thus not constant, but increases with increasing velocity. This phenomenon can be detected experimentally in the beta-radiation of radioactive substances, where changes in mass at high velocities in the particles of these rays—corresponding exactly to Lorentzian prediction—have been determined.

Lorentz's work forms the foundation for Einstein's theories. Lorentz recognizes the idea of an ether at rest, and consequently can speak of absolute motion relative to this. He says furthermore that it would be impossible for an observer in motion and for an observer at rest to determine who of the two was moving and who was at rest. Einstein raises this to a generally valid principle, and postulates that all natural phenomena occur in accordance with identical laws in all systems at rest and systems in constant motion, such that it will be impossible for two observers ever in any way to determine who of them is in motion and who is at rest. This is Einstein's *special relativity principle*. Einstein now goes a step further than Lorentz and says that if it is impossible to determine our absolute motion relative to the ether, then it is indeed unnatural to maintain the idea of an ether in the material sense, and one should ascribe to the word ether only the characteristic in space that it is able to transmit electromagnetic waves. For Einstein then, no such thing as an ether exists, and consequently no such thing as absolute motion either. All motion is relative.

Einstein formulated the laws governing natural phenomena mathematically, and pointed out that the laws of classical mechanics are limit cases of the more generally valid laws established on the basis of the relativity principle.

In 1915, Einstein established his *general relativity principle*, which says that all natural phenomena occur in the same way in all systems, whether they are in accelerated motion or not. This involves gravity along with the electromagnetic interaction.

Here, the example of the elevator comes into play once again, but in a slightly different way. If a physicist is locked in a room, he will not be in a position to determine whether the force that can make the various objects fall to the floor is due to gravity, or to the elevator's traveling up into space in accelerated motion. In the so-called *principle of equivalence*, Einstein says that all natural processes that are affected by gravity occur in the same way they would occur in a space where there was no gravity, and where the whole system was in accelerated motion.

On the basis of the principle of equivalence, Einstein could partly explain, partly predict a series of phenomena that had previously been very puzzling. According to the first Keplerian law of planetary motion, and deducible from Newton's law of gravity, the planets move around the sun in elliptical orbits; but the law was not strictly adhered to, although the expected divergences were so small that they were actually undetectable, with the exception of one case, namely, the orbit of Mercury. It had been observed that the ellipse corresponding to Mercury's orbit precessed slightly more (41 seconds of an arc in the course of a century) than predicted by calculations from classical mechanics. This discrepancy was explicable only by rather implausible hypotheses within classical mechanics; but with the theory of general relativity, it became clear that all the planetary orbits should display similar behavior, only to such a tiny extent that it is in fact immeasurable—except for the case of Mercury. Einstein's calculations gave a value for the case of Mercury corresponding precisely to that found by the astronomers.

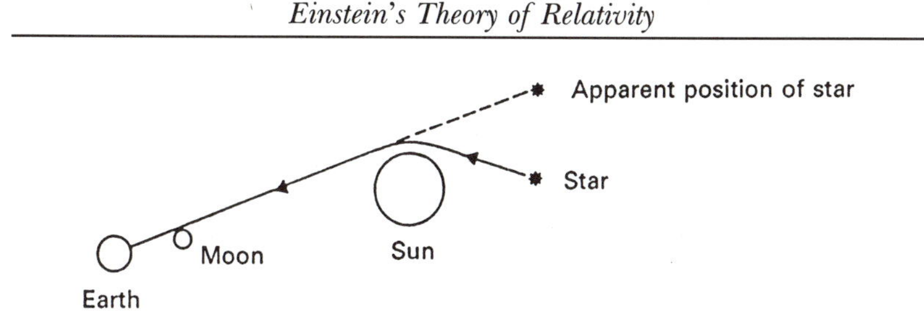

Figure 5.5. Observation of a star under a solar eclipse.

An astronomical prediction made by Einstein demonstrates how far-reaching the principle of equivalence is. Based on this principle, Einstein could prove that light behaved as if it had a certain mass. The prediction was corroborated in the most elegant way during the solar eclipse of 1919. Already in the year 1900, the Russian Lebedev had demonstrated experimentally that the light beams striking the Earth each second is equivalent to a mass of about 1500 kg, a figure fully in agreement with the light pressure Maxwell calculated. When the solar eclipse occurred in 1919, two large expeditions were sent out whose most important assignment was to show that a light beam passing near the sun is affected by its gravity and deflected slightly. It was proved true because the stars observed in the vicinity of the sun were seen away from the places they normally occupied (when the sun was on the other side of the Earth, and far away from the line of sight to the stars); see Fig. 5.5.

Finally, Einstein's theory has in a critical way changed our understanding of concepts such as time, simultaneity, space, mass, and energy. To illustrate the relativity of simultaneity, we use the example of a moving train, on which there is a car with a door at each end and a light source exactly in the middle of the car. When light signals from the light source strike the doors, they open. Light signals are now emitted simultaneously from the source towards the two ends. A person in the train placed at the light source will then see the two doors opening simultaneously, whereas a person outside the train will see the forward door open later than the rear door, since the light seen from outside has traveled a longer distance from the point of its emission from the light source to the point of its striking the forward door, than to its striking the rear door. Events that seem simultaneous in one frame of reference need not, in other words, be simultaneous in another frame of reference—which again implies that time depends on the observer's state of motion.

It is odd that to arrive at simple laws of nature, Einstein was forced to employ concepts of space that deviate profoundly from Euclidian geometry. He thought that space is finite, but without boundaries, such as a three-dimensional sphere, or some other closed three-dimensional manifold. In recent years, much work has been done on the hypothesis that space is probably finite, but that it is continuously expanding in the wake of "the big bang."

3. Minkowski Space-Time and the Special Theory of Relativity

We turn now to the mathematics underlying the theory of relativity. Even though we shall not go into details here, the geometric roots of relativity theory can be found in the non-Euclidian geometry* developed at the beginning of the 19th

* The geometry in the non-Euclidian plane satisfies all the axioms known from the geometry in the Euclidian plane except the parallel axiom, which says that through every point in the plane one and only one line can pass parallel to a given line.

century by Janos Bolyai (1802–1860), Nikolai Ivanovich Lobachevsky (1793–1856), and Karl Friedrich Gauss (1777–1855). The appearance of this non-Euclidian geometry raised the question of which geometry best fit physical observations. This question stimulated Georg Friedrich Bernhard Riemann (1826–1866), a student of Gauss, to develop another type of geometry, known today as Riemannian geometry, which added further weight to the idea that Euclidian geometry does not give the best possible description of the physical world. With a little twist to his geometry, Riemann might well have developed the theory of general relativity before Einstein. Also of great importance to Einstein was the development of tensor analysis, chiefly due to Elwin Bruno Christoffel (1829–1900) and Tullio Levi-Civita (1873–1941).

A fundamental concept in relativity theory is the concept of an *event*. An event is an idealized occurrence in the physical world that has extension in neither space nor time. It could be the snapping of one's fingers or the explosion of a firecracker. Let M denote the set of all possible events in the universe, those that have occurred in the past as well as those occurring now and those to occur in the future. This set of events M forms the basic set in a *space-time*. The problem now is to equip M with a mathematical structure that captures significant physical relations.

First, we consider the special theory of relativity. In this theory, space and time are combined in a 4-dimensional picture of the world called the *Minkowski space-time*. Hermann Minkowski (1864–1909) was a Russian-born mathematician who had lectured to Einstein while the latter was a student at the Polytechnic Institute in Zürich, and it was Minkowski who proposed the geometric space-time viewpoint that formed the foundation for Einstein's special theory of relativity.

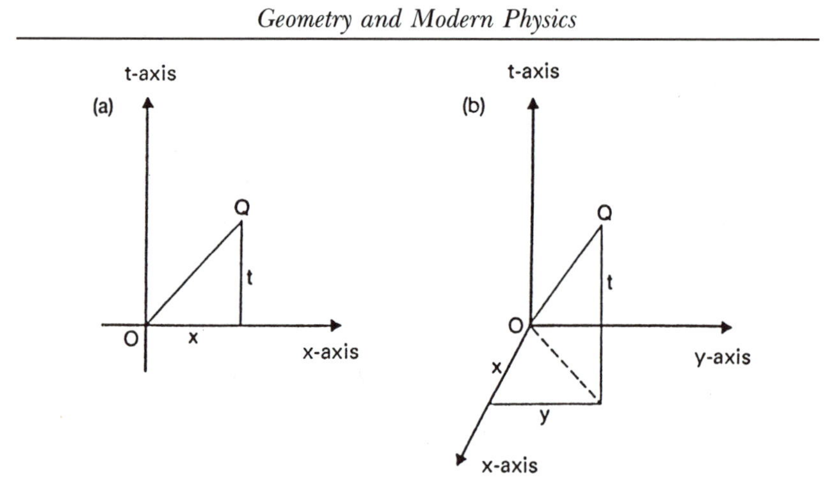

Figure 5.6. (a) The points Q of a plane described by the rectangular coordinates t and x out from the origin O. This plane can be identified with the two-dimensional section in M, determined by $y = z = 0$. (b) The points Q of three-dimensional space described by the rectangular coordinates t, x, and y out from the origin O. This space can be identified with the three-dimensional section in M, determined by $z = 0$.

Minkowski space-time can properly be described by four global coordinates, t, x, y, z, where t is a time coordinate, and x, y, z are space coordinates. The basic set M in Minkowski space-time is identified hereby with the four-dimensional real number space $\mathbb{R}^4$. To form an image of M, it is useful to look at two- and three-dimensional sections in M. In Fig. 5.6, we show the two-dimensional section in M corresponding to $y = z = 0$, and the three-dimensional section in M corresponding to $z = 0$. We identify these sections, respectively, with the two-dimensional plane and the three-dimensional space with the aid of ordinary rectangular coordinate systems. In general, it is advantageous to look at three-dimensional sections in M that

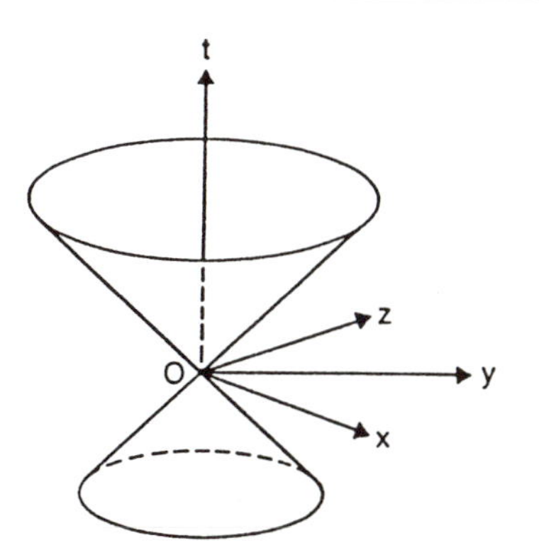

Figure 5.7. Light cone at the origin O in Minkowski space time.

contain the time axis (the t-axis); that is, we represent the three spatial coordinates x, y, z by a plane as in Fig. 5.7.

It simplifies matters somewhat if, in the following, we rescale the time axis by the speed of light, usually denoted c, so that the length $c \cdot t$ is written simply t. With this convention, we then define the *light cone* in the origin (the zero point) O for the Minkowski space time M by the equation

$$t^2 = x^2 + y^2 + z^2.$$

In the three-dimensional section in Minkowski space-time in Fig. 5.7, where the three spatial coordinates are represented by a plane, the light cone consists of an ordinary cone that forms a slope of 45° to the t axis. The light cone comprises two parts, namely, the *future light cone* fixed at $t > 0$, which represents the history of a spherical light wave that is moving outwards from a light flash at time $t = 0$ in the origin O, and the

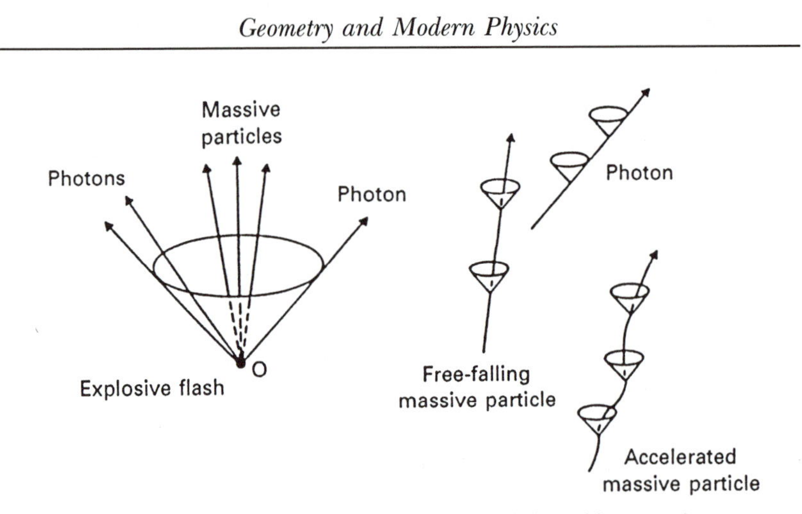

Figure 5.8. World lines of particles in Minkowski space time.

past light cone, fixed at $t < 0$, which represents the history of a spherical light wave that is moving inwards towards the origin 0, where it arrives at time $t = 0$.

We can think of light as if it propagates as particles, the so-called *photons*. The generators of the light cone, by which we mean the straight lines on the cone through the origin O, thus represent the history of the individual photons from the light flash. The same picture holds also for other types of massless particles, since all such particles move at the speed of light. A massive particle, on the other hand, must always move at a speed less than that of light, so the history of a free massive particle sent out in the same explosive flash that produces photons will be a straight line extending from the origin O and lying inside the future light cone. See Fig. 5.8.

Every other particle in free, unaccelerated motion is likewise represented by a line in Minkowski space. For a massless particle, the line will appear as a straight line, that forms a slope of 45° in relation to the t axis; for a massive particle, the slope will be less than 45°. A massive particle not in free motion is described by a curve called the *world line* of the particle, for which the tangents everywhere have a slope less than 45° in relation to the t-axis; that is, they lie everywhere within the local lightcone. See Fig. 5.8. Such a world line is said to be *time-like*.

In every kind of geometry, the measurement of length is a crucial concept, and we shall therefore now equip Minkowski space with a measurement of length; mathematically, we speak of a *metric*. It may seem obvious that the time-axis plays a very special role, and that this must be reflected in the measurement of length, but it is perhaps precisely in the choice of the measurement of length that genius resides. Let us first, however, consider the Euclidian measurement of length.

By choosing a rectangular coordinate system, the Euclidian plane can be described by the coordinates t and x as in Fig. 5.6a. The Euclidian distance $|OQ|_e$ from the origin O to a point Q with coordinates t and x is then given by the formula $|OQ|_e^2 = t^2 + x^2$. This follows immediately from the Pythagorean theorem, which says that the square of the hypotenuse of a right triangle is equal to the sum of the squares of the two sides. If we similarly describe a point Q in Euclidian space by rectangular coordinates t, x, and y, as in Fig. 5.6b, then the Euclidian distance $|OQ|_e$ from the origin O to Q is given by the formula $|OQ|_e^2 = t^2 + x^2 + y^2$.

We are now ready to consider Minkowski space. We define the Minkowski distance $|OQ|_m$ from origin O to a point Q, with coordinates t, x, y, z, by the formula

$$|OQ|_m^2 = t^2 - x^2 - y^2 - z^2.$$

The corresponding Euclidian distance is obtained by substituting plus signs for the minus signs in the formula. In the spatial coordinates, the measurement of length is Euclidian (except for the sign), while the time coordinate has the opposite measurement of length. We say that the geometry of Minkowski space is pseudo-Euclidian. Note that $|OQ|_m^2 = 0$ if Q lies on the light cone in O, and that $|OQ|_m^2 > 0$ if Q lies within the light cone in O. The meaning of the "distance" $|OQ|_m$ is that when Q is within the light cone in O, it measures the time interval that occurs between the events O and Q, measured by a clock whose world line is the straight line segment OQ. Therefore $|OQ|_m$ is also said to be the *proper time* that passes in traveling from O to Q along the line segment OQ.

There is a corresponding result for world lines that do not pass through O. If P has the coordinates t', x', y', z', then the Minkowski distance $|PQ|_m$ is given by the formula

$$|PQ|_m^2 = (t - t')^2 - (x - x')^2 - (y - y')^2 - (z - z')^2.$$

We see that $|PQ|_m^2 > 0$ if Q lies within the light-cone in P, and that $|PQ|_m$ in this case measures the proper time interval that elapses on a clock in free unaccelerated motion from P to Q.

The fact that the measurement of time takes place according to this kind of expression, and not according to the Newtonian "absolute" time difference $t - t'$, is the key to the special theory of relativity. Time is the "distance of travel" measured according to Minkowski geometry; and just like ordinary measurement of distance of travel in Euclidian geometry, the measurement of time along a time-like curve (a world line) is dependent on the chosen curve between two events. If P and Q are two points in Minkowski space connected to each other by

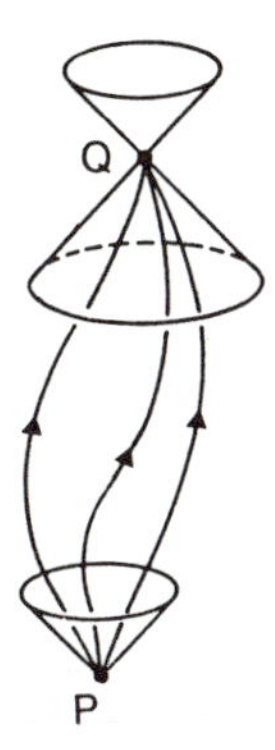

Figure 5.9.

several time-like curves (see Fig. 5.9), the proper time interval between P and Q will in general be different along the different curves.

Although the foregoing discussion may seem odd at first, it must be emphasized that there is overwhelming experimental support for such a non-absolute time measurement; for example, in the measurement of decay of particles in the cosmic radiation created in the upper atmosphere, in precise measurements of time in aircrafts, and in the behavior of particles in high-energy accelerators. We are altogether familiar with the notion that ordinary Euclidian distance of travel is a path-dependent concept. In contrast to this, our intuition concerning time is built up on the basis of experiences where the path-dependence is utterly minimal, because ordinary speeds are so much lower than the speed of light, so that the term $(t - t')^2$ is entirely dominant in the formula for the Minkowski distance $|PO|_m^2$.

In Euclidian geometry, the straight line PQ is that curve along which the distance of travel is minimal. A curious inversion of this relation occurs in Minkowski geometry, in that the

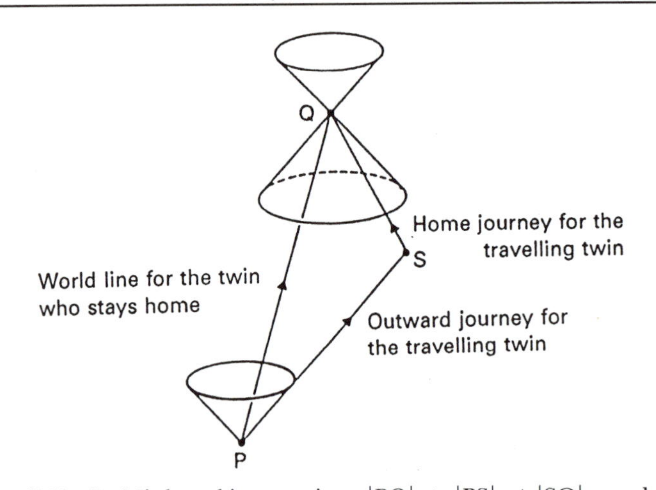

Figure 5.10. In Minkowski space-time, $|PQ|_m > |PS|_m + |SQ|_m$, and this gives rise to the twin paradox.

straight unaccelerated world line from P to Q is that time-like curve from P to Q along which the Minkowski distance of travel is maximal, or in other words, that curve where the proper time interval is greatest. This is related to the so-called *twin paradox* in the special theory of relativity, which is illustrated in Fig. 5.10. Two twin sisters start from P, one of them remaining unaccelerated on the Earth, while the other undertakes a round-trip journey to a distant planet S at a velocity close to the speed of light. When the twins are reunited at Q, it turns out that the traveling twin has aged less than the twin who stayed home. The explanation is that the passage of proper time experienced by the twin remaining at home during her sister's trip is exactly $|PQ|_m$, and this time lapse has to be greater than the proper time lapse measured along the traveler's worldline, which after all diverges from the straight line.

4. Curvature and Gravitation: The General Theory of Relativity

Minkowski space is inadequate for incorporating gravitation. Einstein solved this problem by replacing Minkowski space with a curved space-time manifold. This leads to the general theory of relativity, which we shall now briefly describe. This section will probably require some patience on the part of the reader.

Let M once more denote the set of all possible events in the universe, those that have occurred in the past as well as those occurring now and those that are to occur in the future. The idea is, as before, to introduce additional structure onto M to account for idealizations of different experiments one could imagine performed in the physical world.

Let us consider a subset of M—for example, all the events that take place in a fixed room during a given time interval. There are different ways in which one can identify this subset with a subset of the four-dimensional real number space $\mathbb{R}^4$ consisting of all ordered sets (t, x, y, z) of four real numbers, which is the basic set in Minkowski space. For example, one could imagine the room filled from floor to ceiling with point-shaped clocks. Each clock is thus associated with three numbers—e.g., the distance from the forward wall, from one side-wall, and from the floor—determining the clock's position in the room (Fig. 5.11). When we select an event in the subset, the snapping of a pair of fingers in the room, say, we read the time on the clock and note the position of the clock at the point

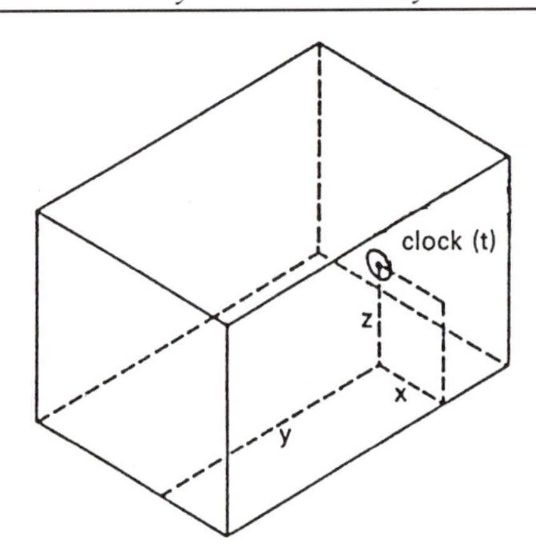

Figure 5.11.

where the snapping occurred to obtain the four coordinates of the event.

Let us assume now that constructions of this type give smoothly overlapping local charts on M. (See Chapter 2.) These remarks suggest that it makes sense to furnish M with the structure of a smooth four-dimensional manifold.

We regard the events as the fundamental entities in the physical world, and therefore adopt the position that more complicated phenomena ought to be described in terms of these fundamental entities, when at all possible. Several examples of this point of view follow.

We wish to describe the history of a particle by means of M. Consider the set of all events occurring where the particle is. Our intuitive picture of a particle tells us that this set must be a one-parameter family of points in M, since an event can occur only where the particle is, but at any point in time. Therefore, it is reasonable as a mathematical model for a particle (a physical thing) to use a curve in M. In a similar manner, the physical thing "a segment of a thread" will be idealized and described mathematically as a two-dimensional surface segment in M, and the physical thing "a piece of paper" correspondingly as a three-dimensional submanifold in M.

As another example, let us consider the conventional picture of the world, in which "space" and "time" are more or less separate things. We imagine once more that our experimental room from before is filled with clocks. Associate a real number with every event in this room, namely, the time on the clock where the event occurs. Thereby, we associate a number with events, corresponding in the idealized model to our defining a real function t on M. For every fixed real number t_0, the subset of M on which $t = t_0$ represents the entire space at the time t_0 (and this will be a three-dimensional submanifold of M). Two events will be described as occurring at the same point in time if their t-values are the same. It turns out, however, that there is no physical construction that yields a preferred t-function; the t resulting from a construction like the one just suggested depends on the details of how the clocks' positions are determined, and how they are synchronized, etc. (*cf.* the relativity of simultaneity discussed in Section 5.2). For this reason, one is apparently forced into a model involving a four-dimensional manifold M, and not just a family of three-dimensional manifolds labeled by t. Mathematically, one would say that a time-axis cannot be split off from M. To illustrate this situation, we show in Fig. 5.12 some two-dimensional examples which one can think of as two-dimensional sections in M.

We shall now argue that it makes sense to furnish M with a (pseudo-Euclidian) measurement of length for the tangent vectors to M; mathematically, we speak again of a *metric*. By a *tangent vector* to M in a point P, we mean here the velocity vector in P for a "particle" passing through P. The set of

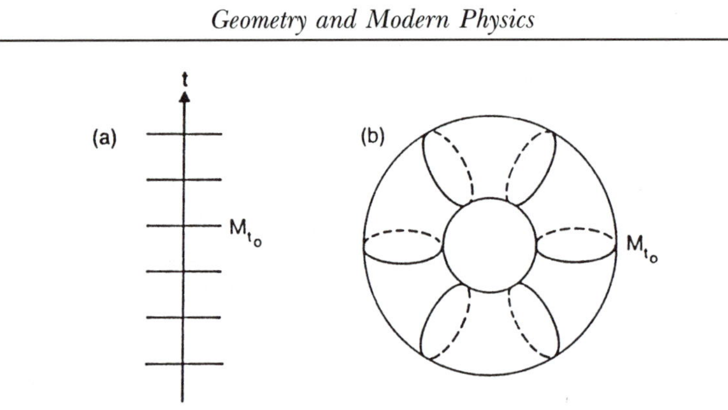

Figure 5.12. (a) A two-dimensional example with a split off time axis. (b) A two-dimensional example in which a time axis cannot be split off.

tangent vectors to M in P is a four-dimensional linear space called the *tangent space* to M at P. If a local chart around P is chosen, the tangent space can naturally be identified with the four-dimensional real number space $\mathbb{R}^4$. In Fig. 5.13, we have shown how the corresponding identification takes place in dimension 2. A measurement of length at the tangent vectors to M will automatically give a corresponding measurement of length for curves on M. This is immediate when we think of tangent vectors as velocity vectors for "particles" that follow the curves, for if we know the velocity of a "particle" along a curve at every point in time, then we can calculate the distance covered (by a mathematical process called *integration*).

Let us assume that it is feasible to build clocks that for all practical purposes are identical and whose internal workings are affected neither by external factors nor by the clocks' past history. Every clock has a face displaying at every point in time a number that is the time according to that clock. The time is

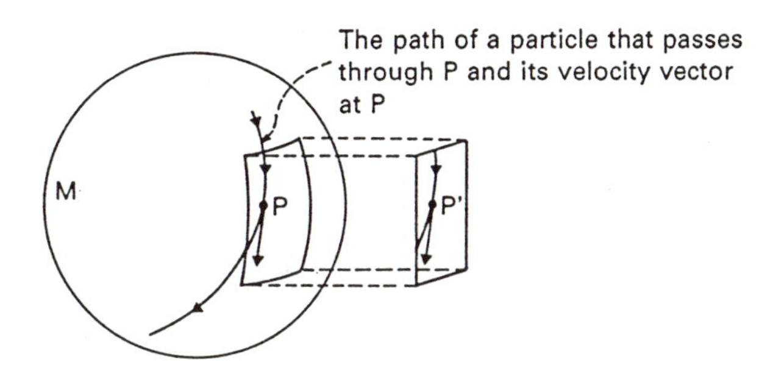

Figure 5.13. In dimension 2, one can think of a local chart around P as a patch glued to a surface. When the patch is glued on, its tangent space at P′ (that is, $\mathbb{R}^2$) is identified with the tangent space to the surface at P.

to be measured here as proceeding continuously, as in a clock incorporating a calendar—not just periodically, in 24-hour periods. Each of these (point-shaped) clocks are thus represented in M by a curve. The time read on such a clock associates a real number with every event occuring in the immediate vicinity of the clock. Within the model, therefore, we replace reading the time from the clock with a parameter t along the curve described by the clock in M. In mathematics, such a curve is called a *parameterized curve*. Seen mathematically then, a clock is a parameterized curve γ in M, where the parameter represents the time.

The special theory of relativity is physically so well-documented that we must insist that the mathematical model M of space-time locally is as in the special theory of relativity; that is, on a small scale it resembles Minkowski space.

Now consider two fixed events, P and Q, and consider a clock γ that passes through P. From the special theory of

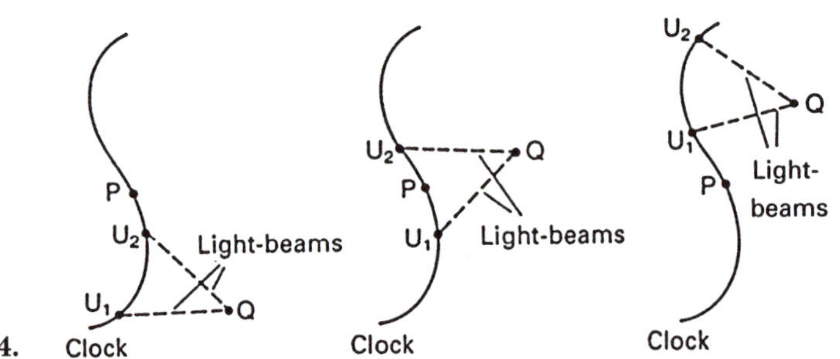

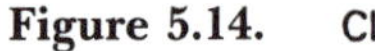

Figure 5.14.

relativity (geometry in Minkowski space), it follows that if P and Q lie sufficiently close to one another, there will be precisely two light beams, each of which has the characteristic that it meets both the clock and the event Q. (See Figure 5.14.) From a physical point of view, a light beam is sent out from the clock at exactly the right moment (event U_1 in the figure), and in just the right direction so that it arrives at event Q. The other light beam is sent out from Q in the right direction so that it returns to the clock (event U_2 in the figure). Now set

$$d(P, Q, \gamma) = (t(P) - t(U_2)) \cdot (t(P) - t(U_1)),$$

where $t(R)$ is the time registered on a clock in the immediate vicinity of an event R. On the basis of the special theory of relativity (geometry in Minkowski space) it can further be concluded that when Q approaches P (with γ and P fixed), then $d(P, Q, \gamma)$ becomes independent of γ and quadratic in the deviation of Q from P, as the Minkowski distance $|PQ|_m^2$ in Minkowski space.

Next, fix P and γ, and let Q be a "particle" passing through P. By a mathematical process (for the initiated, two times differentiation of $d(P, Q, \gamma)$ along the trace for Q expressed in local coordinates on M around P), it is possible to define the square of a measurement of length of the velocity vector for "the particle" Q at the moment it passes through P. Thereby, the sought-after pseudo-Euclidian measurement of length (the metric) is determined for the tangent vectors to M in P. Since the construction agrees locally with the geometry in Minkowski space, there must in every tangent space exist a *light cone*, within which vectors have positive lengths and are called *time-like vectors*, while vectors outside have negative lengths and are called *space-like vectors*.

Our mathematical model of the universe, therefore, consists of a four-dimensional manifold M with a metric on the tangent vectors to M, such that in every tangent space there is a light cone, within which lengths of vectors are to be reckoned positive, and outside of which lengths of vectors are to be reckoned negative. In "the direction of the light cone," lengths are positive; in the "three spatial directions," negative. Such a model is called a *space-time manifold*.

We must introduce one more important connection between experiences from the physical world and the mathematical model. Imagine constructing an *accelerometer* consisting of a small rigid frame within which a mass is suspended from springs. (See Fig. 5.15.) At an arbitrary point in time, the mass may assume a position with respect to the frame that deviates from its normal position in the center of the frame. For example, the mass will be drawn downwards if the accelerometer is placed on the Earth, and if we accelerate the object, it will be displaced towards one of the ends. The center of the frame can be described by a curve γ in M, while the mass will be represented by another curve γ', such that if the mass is to remain at the center of the frame, it corresponds mathematically to the coincidence of the two curves γ and γ'. We want

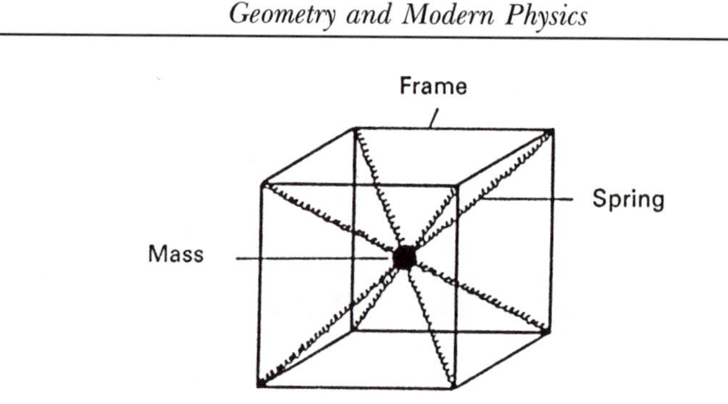

Figure 5.15. Accelerometer.

to go to the limit by having a vanishingly small accelerometer, that is, one for which the lengths of the edges of the frame approach zero. In this limit, we can replace γ' with a vector field **A** along γ, which at every point of γ measures the deviation of γ' from γ in the plane orthogonal (perpendicular) to γ (in the Minkowski geometry). This vector field represents the acceleration of the curve γ.

In an idealized description then, the accelerometer is a curve γ in M, where in every point a vector **A** is specified orthogonal to the tangent vector to the curve γ. See Fig. 5.16. The vector field is called the *acceleration* of the curve, and it describes the change in velocity experienced by an object whose motion is determined by the curve. In general relativity theory, the generalized measurement of length on M is arranged just so that the acceleration of the curve γ will come to correspond geometrically to the *curvature* of the curve in relation to M. We take the view that a vector field **A** different from the zero field shall be interpreted as the effect of an "external force." We would therefore expect a field different from the zero-field for a charged accelerometer in an electric field, and for an

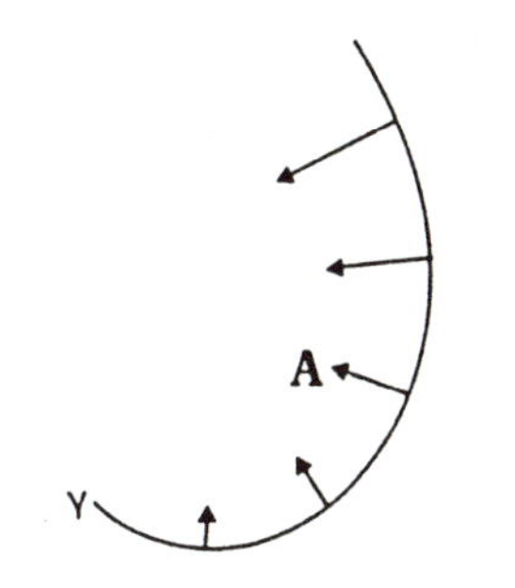

Figure 5.16. Acceleration **A** of a curve γ.

accelerometer placed on a table. In the first instance, the external force is an electrical force, while in the second, it is the reaction from the table on the accelerometer. In particular, a free particle (that is, a particle not affected by any external force) must be described by a curve with zero acceleration, or what amounts to the same thing, a curve with zero curvature. In geometry, such a curve is called a *geodesic curve*. In the Euclidian plane, the geodesic curves are the straight lines. On a sphere, they are the great-circles.

We did not include a "gravity" among the external forces in the preceding discussion. The reason is that we know of no way to measure anything we could call a "gravity" (*cf.* the equivalence principle). If we place an accelerometer on a table on the surface of the Earth, the acceleration is accounted for by the upward-directed reaction from the table; if we drop an accelerometer from a point above the Earth's surface, it will display zero acceleration. The reason for this is that gravitation affects the mass inside the accelerometer in the same way it affects the frame itself, so that the instrument is unable to measure a "gravity."

This concludes the so-called *kinematic* part of the general theory of relativity, that is, the theory of motion. We have, however, not yet described a proper physical theory, for there we also have to describe *the dynamics*, that is, the changes of systems in the physical theory. To this end, one employs as a rule (partial) *differential equations*, that is, equations that relate mathematical quantities with associated *differential quantities*, which describe changes. As a rule, changes in a physical system are caused by a field, and for the mathematical treatment it is important here that a field can often be described by a *potential*, which expresses the energy in the field; in other words, its ability to carry out work, i.e. its ability to change the state of a system.

As a model for the general theory of relativity in the preceding, we have introduced a manifold M, by means of which we can represent particles, light beams, etc., and a metric, by which we can describe certain physical experiences, such as the passing of time and acceleration. Even if the apparatus developed here already entails certain predictions—such as that two clocks traveling from event P to event Q along two different paths may well display different times—there is nevertheless too much freedom in the metric. In order to make the detailed quantitative predictions characteristic of a physical theory, the metric has to be connected in one way or another with the distribution of matter in the universe. In Newton's theory of gravitation, there is a gravity potential that is related by a differential equation to the density of mass (Poisson's equation). The potential in Newton's theory corresponds to the metric in the general theory of relativity, for both determine the motion of a test particle. There is also the need for an equation in general relativity theory corresponding to the equation in Newton's theory of gravitation. This equation was found by Einstein, and it relates curvatures derived from the metric with a so-called *energy-momentum tensor* on M, corresponding to the density of mass in Newton's theory.

What exactly the energy-momentum tensor is depends on the details of what matter is present. Each type of matter (liquids, solids, electromagnetic fields) is described within the model by suitable tensor fields* on M, subject to appropriate differential equations.

By way of conclusion then, we may briefly review the mathematical layout of the general theory of relativity as follows. The underlying space is a four-dimensional smooth manifold M. On this manifold, there is a metric that gives rise to a light cone in every tangent space for the manifold with one "temporal" and three "spatial" dimensions. Physical effects on particles, etc., are described by means of the metric. There is in addition an energy-momentum tensor on M, which describes the distribution of mass. The metric and the energy-momentum tensor are related by a suitable equation called *Einstein's equation*. Gravitational effects arise because Einstein's equation dictates that the presence of matter introduces curvature, and this metrical curvature then exerts influence on other objects in the universe.

One can reasonably pose the question whether it is necessary at all to assume that space-time is curved. The answer is

* A *tensor field* on M associates with every point in M a geometric quantity that in a local chart around the point takes the form of a matrix of numbers; a tensor field can be thought of as a kind of generalized vector field.

that we can in fact detect small deviations from the flat Minkowski space, due to the presence of a gravitational field. Thus, in 1960, Pound and Rebka carried out an experiment demonstrating that the relation between the speed of a clock at the top and at the bottom of a 22.6-meter-high tower is not 1, but 1.0000000000000025, and it is just this ratio that is predicted on the basis of the general theory of relativity. Since clock speeds measure the geometry of space-time, we can take this as a direct measure of the deviation of the geometry of space-time from the flat Minkowski space. There are indeed many other, more decisive tests of the theory's validity, and the general theory of relativity must be regarded as being quite well-documented. For example, we get a rather direct confirmation that the null-cone (the light cone) is not of pure Minkowski type through the deflection of light in the gravitational field of the sun, since light beams *in vacuo* must follow the exact null-cone. As mentioned, the deflection of light was observed by two expeditions sent out to research the solar eclipse in 1919. More precise measurements with radar signals have been carried out more recently, and are in very clear agreement with the predictions of the general theory of relativity.

General relativity is a fruitful marriage between (differential) geometry and the physical theories of gravitation. Ever since this marriage was entered into at the beginning of this century through works of Einstein, the great German mathematician David Hilbert (1866–1943), and many others, it has been a fertile alliance for both fields. On the physical side, its success can be measured by the very exact agreement of Einstein's classic theory with all the gravitational experiments and astrophysical observations carried out as cited before, and with the theory's own predictions of new and interesting effects, for some of which there is now solid support. As examples, we can mention black holes, the expansion of the universe, gravitational waves, and gravitational lenses. From a mathematical point of view, the theory of general relativity has stimulated a fruitful study of pseudo-Riemannian as well as Riemannian geometry, and it has been the impetus for important work in complex geometry, in topology, and in the study of partial differential equations.

One of the most important recent advances is the proof of the so-called Positive Mass Theorem, proved in 1979 by the mathematicians Richard Schoen and S.-T. Yau, and independently also by the mathematical physicist Edward Witten. The theorem clarifies a classic conjecture in the theory of general relativity, according to which the total mass of an asymptotic flat (its curvature approaching zero at infinity) gravitational field is always positive. It was the validity of this conjecture for a general physical system that Schoen and Yau, and independently Witten, proved in 1979. They have all received the highest forms of recognition; Yau was awarded the Fields Medal in 1982 and Witten in 1990.

In the theory of relativity, mass and energy are equivalent quantities, related by Einstein's famous equation $E = m \cdot c^2$, where E is the energy, m is the mass, and c is the speed of light. Since gravity is attractive, it gives rise to a negative energy component in the total energy, and one might therefore imagine for example that a star in a gravitational collapse to a black hole would pass through a state with a dominating gravitational energy and thus negative mass. It is conjectured, however, that this cannot take place, since everything suggests that an event horizon rises around the collapsing star, and a black hole is formed before the negative gravitational energy exceeds the star's positive component of matter. This conjecture has yet to be proved, however.

5. The Physics of Elementary Particles

Physics is the study of the dynamical world—interactions are what interest physicists. In the last three sections of this chapter, we shall survey geometry in relation to the physics of elementary particles. The dynamical world on the atomic level is quantum-mechanical, a fact that cannot be denied. Quantum mechanics associates with events the probabilities of their occurring in the real world. Whereas in classical mechanics one would say, "The sun will rise tomorrow morning at 7 o'clock," in quantum mechanics one would say, "The probability that the sun will rise tomorrow morning at 7 o'clock is almost, but not quite, 100%." In the macroscopic world, the difference between quantum mechanics and classical mechanics is quite irrelevant; but in the sub-microscopic world, classical mechanics breaks down completely, while quantum mechanics remains uncontradicted.

In *quantum mechanics*, established in the 1920s by Niels Bohr (1885–1962), Werner K. Heisenberg (1901–1976), and Erwin Schrödinger (1887–1961), among others, a particle or any other physical object is described by a probability field, which expresses the particle's or the object's tendency to be in a given state. Mathematically, a particle is described by a *wave function*, which to every point x in the space-time manifold M associates a vector $\psi(x)$, whose direction represents the *phase*, and whose magnitude represents the *amplitude* of the oscillation. The square of the length of $\psi(x)$ describes the probability that the particle is close to the point x in the space-time manifold.

Quantum mechanics yields an entirely new view of the world with a dualism between the description of waves and the description of particles. For classical electrodynamics, this leads to a quantizing of the electromagnetic field into so-called

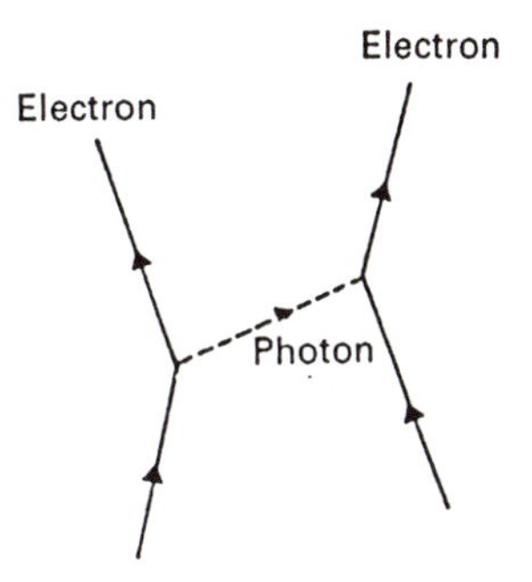

Figure 5.17. Mutual repulsion of two electrons by the exchange of a photon in a so-called Feynman diagram.

photons. These appear as particle manifestations of the electromagnetic waves. The theory describing this is called *quantum electrodynamics*, and it incorporates both relativity theory and quantum theory. It was the first relativistic quantum theory in physics, and it remains the most successful union of the two theories.

It is not just the electromagnetic field that is quantized. Every field can be quantized, and we can therefore speak generally of a *quantum field*, that is, a field that can take the form of quanta or particles. The gravitational field is quantized into so-called *gravitons*. The interaction between atomic particles is also a quantum field described by fictitious particles, so-called *virtual* particles. The description occurs in diagrams named after the famous American physicist Richard P. Feynman (1918–1988). Thus, two electrons interact with each other by exchanging photons (see Fig. 5.17), and the mutual repulsion is an expression of the collective effect of multiple photon exchanges.

In *quantum field theory*, the classical distinction between material particles and the surrounding space is abolished completely. The quantum field is a fundamental physical quantity; it is a continuous medium, present everywhere in space. Particles are just local densities of the field—concentrations of energy that come, and then go, losing their individual character and dissolving in the underlying field.

The concept *elementary particle* seems to be strongly time-dependent, and it has changed under the influence of the knowledge obtained at any time. In this century, we have seen a procession of more and more elementary particles—from molecules to atoms, from atoms to electrons and nucleons, from nucleons to positrons and neutrons—which are assemblies of quarks, etc. Elementary particle is the collective term comprising the building blocks that at any time are regarded as the fundamental ones. Whenever the physicists have had greater energy sources at their disposal to smash atoms, they have discovered new elementary particles.

Elementary particles are divided into leptons and hadrons. The *leptons* are particles that can interact with other particles via the electromagnetic and the weak interaction. *Hadrons* are particles that can in addition interact with other particles via the strong interaction. Today, six leptons are known, among them the electron. Of the hadrons, in the last few decades more than 200 have been identified. These have in turn been divided up into two categories: mesons and baryons. *Mesons* have integer spin (see later in this section), and *baryons* have half-integer spin. Examples of mesons are the pions, which transmit a strong interaction. The best-known baryons are the proton and the neutron, designated jointly as nucleons. All other baryons are heavier than the proton and will sooner or later decay to protons in the emission of mesons and leptons. One can therefore think of baryons as excited protons.

One of the most fruitful concepts in elementary particle physics is the concept of symmetry. Global space-time symmetries, such as translations, rotations, etc., have been known and used for many years; these symmetries are called *external symmetries*. Beyond global symmetries, which almost always appear in physical theories, it is possible to have local symmetries, where the convention of symmetry can be determined independently at every point of space and at every point in time. Although the word "local" might suggest something less significant than global symmetry, the requirements for local symmetries are in fact what make the strongest requirements by far in the constuction of a physical theory. A global symmetry for a physical law states that the law remains in force when the same transformation (in the given symmetry class) is carried out in all places at the same time. In order that a local symmetry for a physical law can be observed, the regularity must retain its validity even when different transformations are employed (in the given symmetry class) in the various points of the space-time manifold. These local symmetries, called *inner symmetries*, play a crucial role for elementary particles. An inner symmetry leaves the equations of motion for the particles unaltered without affecting the space-time variables. An example of such an inner symmetry is *spin*, which is a rotation symmetry of suitable order around the origin in three-dimensional space. A particle with integer spin returns into itself by a full rotation around any particular *spin axis*, whereas a particle with half-integer spin goes into another state with a full rotation around spin axis, and returns to its original state only after a double rotation around the spin axis.

Conditions of symmetry are described mathematically using the concept of a group. By a *group*, we mean a set G of elements furnished with a multiplication (a product operation), which to every pair of elements a and b in G associates a product element $a \cdot b$ in G. The product operation must be associative, that is $(a \cdot b) \cdot c = a \cdot (b \cdot c)$. There shall exist an element 1 in G, which is neutral with respect to multiplication, and every element a in G must have an inverse element a^{-1} in G. These elements are characterized by the arithmetic rules $1 \cdot a = a \cdot 1 = a$ and $a \cdot a^{-1} = a^{-1} \cdot a = 1$. A familiar example of a group is the set of positive real numbers with the usual rule of multiplication. Another example is the group consisting of the numbers 1 and -1 with usual multiplication. In this example, the neutral element is 1, and each of the elements has itself as the inverse element. This group is called the *cyclic group of order 2*. The group is generated also by a rotation of 180° in the plane, since this rotation, when it is performed twice, leaves the points in the plane in their original positions. We say that the rotation has order 2. A third example is the circle, where we conceive every point on the circle as an angle of direction, measured from a fixed half-line out from the center of the circle, and where we employ usual addition of angles as multiplication. (See Fig. 5.18.) The point of intersection of the half-line and the circle is the neutral element, and the inverse element to a point on the circle is the point one reaches by going the opposite angle out from the neutral element. This group is called the *circle group*. A fourth example is the set of rotations in space around a fixed zero-point, with composition of rotations as multiplication. This group is called the *rotation group in three-dimensional space*, and is denoted by SO(3). Another interesting group in elementary particle physics is the group U(2), which is a group of rotations in two-dimensional complex number space.

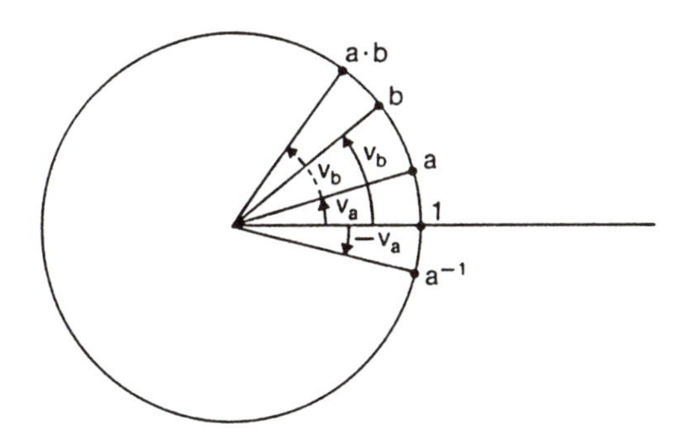

Figure 5.18. Points a and b on the circle correspond to the angles v_a and v_b, respectively. The product element $a \cdot b$ in the circle group thus corresponds to the angle $v_a + v_b$. The inverse element a^{-1} to a corresponds to the angle $-v_a$.

The object of quantum field theory is to a large degree to use the same procedure employed in the description of photons to describe all elementary particles. Where photons stand out as quanta for classical electromagnetic theory, other elementary particles shall stand out by quantizing suitable classical field theories. In the last three decades, gauge theory has emerged as the most promising candidate, and in the Yang–Mills equation we find a generalization of Maxwell's equations (*in vacuo*).

6. Fiber Bundles and Parallel Displacement in Fiber Bundles

The mathematical side of gauge theory is in fact a well-established branch of differential geometry, known as the theory of

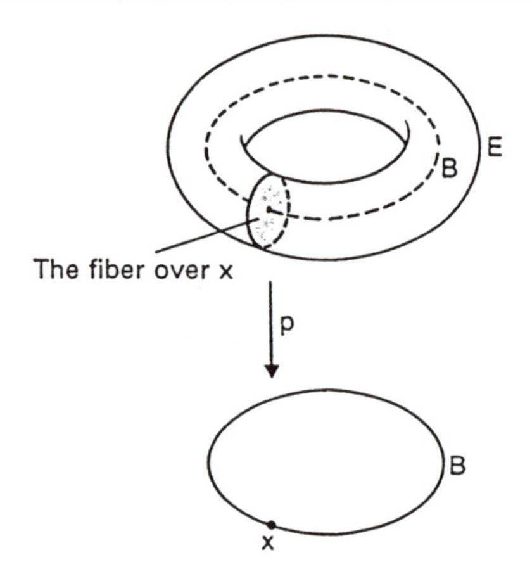

Figure 5.19. The total space E is a massive torus, which by the projection p is projected onto its central circle, which constitutes the base space B in the fiber bundle $p{:}E \to B$.

fiber bundles with parallel displacement. It has much in common with the geometry that provided Einstein with the fundamental mathematical concepts of the general relativity theory. It is by no means an easy theory, but we must attempt to give an impression of it.

A *fiber bundle* is a mathematical structure consisting of two point sets, called the *base space* B and the *total space* E, and a mapping $p\colon E \to B$ called the *projection*, which to each point in E associates a point in B. The point sets will typically be manifolds; for example, the circle or closed surfaces, as in Chapter 2. The relation between the points in the total space and in the base space is often described by thinking of the total space as lying "above" the base space. The set of points in E, which is mapped into the same point $x \in B$, is called the *fiber* over x. See Fig. 5.19.

Some examples will be necessary to illustrate this. The simplest example arises by considering two arbitrary spaces B and F, and forming the product set E, which consists of all ordered point pairs (x, v) of a point x in B and a point v in F. If we now let $p{:}E \to B$ be the projection that to the point (x, v) in E associates the point x in B, we have a fiber bundle with base space B, total space E, projection p, and fiber F over every point in B. This fiber bundle is called the *trivial* bundle with base space B and fiber F. The fiber bundle in Fig. 5.19 is trivial with the circle as base space and the circular disc as fiber.

In the next example of a fiber bundle, we employ the closed surfaces from Chapter 2. Consider the mapping $p\colon S^2 \to \mathbb{R}\mathrm{P}^2$ of the sphere S^2 onto the projective plane $\mathbb{R}\mathrm{P}^2$, which associates with a point on the sphere the corresponding point in the projective plane. A pair of antipodal points on the sphere is hereby mapped into the same point in the projective plane, and conversely just one pair of antipodal points lies over each point in the projective plane. Here then, the fiber consists of two points. In this example, the fiber can be given further mathematical structure, namely, structure such as the cyclic group of order 2, since over suitably small areas of the projective plane it is possible to think of a pair of antipodal points as $+1$ and -1, multiplied together according to the usual rules. A fiber bundle with this structure is called a *principal bundle* with the cyclic group of order 2 as *structure group*.

In the third example of a fiber bundle, we consider the sphere S^2. Regard the sphere as the surface of the Earth, and let us imagine that a person is placed at every point. Any such person can now look around in every direction, perceiving the horizon in the distance as a circle. With every point x in S^2,

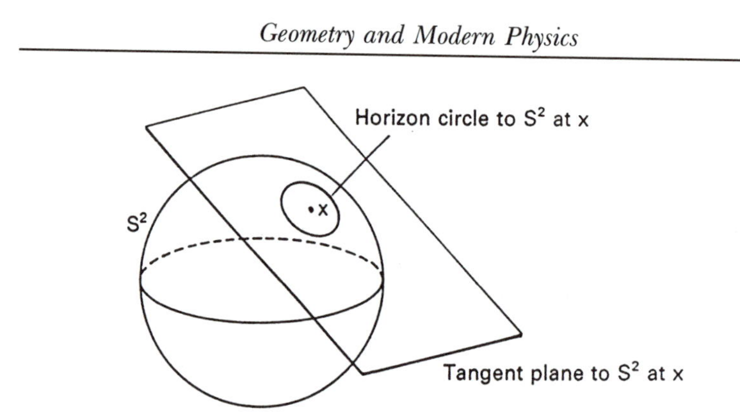

Figure 5.20. The set of all (horizon) circles at points on S^2 constitutes the total space E in a circle bundle over S^2.

therefore, we can associate a (horizon) circle. The set of all these circles constitutes a space E, and there exists a natural projection $p: E \to S^2$ that maps each of these circles into the point with which it is associated. (See Fig. 5.20.) What we have described here is a circle bundle over the sphere. By utilizing the circle's group structure, we obtain in this case a principal bundle with the circle group as structure group. The dimension of E, of course, remains 3, of which two dimensions come from the sphere (the position), and one dimension comes from the circle (the direction). Similar constructions can be carried out with the other closed surfaces as base spaces.

Finally, we must explain what is meant by parallel displacement in a fiber bundle. A *parallel displacement in a fiber bundle* is an operation that, for every path in the base space and every prescribed point in the fiber over the starting point, selects unambiguously in the total space a path beginning in the prescribed point. We must be content here to indicate the idea by considering the circle bundle of (horizon) circles on the sphere

S^2 just introduced. Think once more of the spherical surface as the Earth. A point in the total space E is thus a point on the Earth's surface together with a given direction at the point. From a mathematical point of view, a point in E is nothing else but a tangent vector (of length 1) to the sphere S^2. Given a path on the sphere together with an assignment of a direction (a tangent vector of length 1 to the sphere) at the path's starting point, we can ascribe a direction (a tangent vector of length 1 to the sphere) to every point along the path. We do this as follows. First, we imagine that the path on S^2 and the starting vector are freshly painted and still wet. Next, we place the sphere on a plane such that it rests on the path's starting point. Then we roll the sphere across the plane along the curve on the sphere. A curve will thereby be transferred to the plane, along with a vector in the starting point of the plane curve. We make a parallel displacement of this vector in the usual way along the plane curve, and then roll the plane directional field up on the sphere using the opposite rolling process. The resulting directional field on the sphere is the required "parallel displacement" of the original tangent vector along the path on the sphere. The whole process is shown in Fig. 5.21. We owe the method, known as *Cartan rolling*, to the great French differential geometer Elie Cartan (1869–1951).

7. Gauge Theories and String Theories

Gauge* theory appears for the first time in physics in a 1918 paper by Hermann Weyl in which he attempts to unify general relativity theory and electromagnetism. Weyl had noticed that

* A *gauge* is a device with which to measure something.

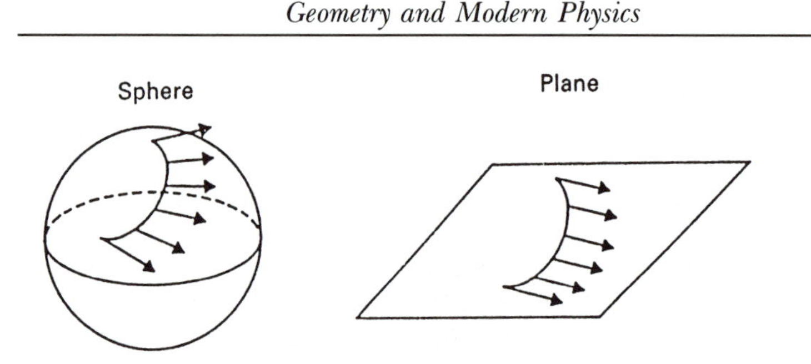

Figure 5.21. A parallel field of tangent vectors to the sphere is rolled out into an ordinary parallel field of vectors in the plane.

Maxwell's equations are invariant with respect to alterations of scale (conformal invariance), and attempted to exploit this fact by interpreting the electromagnetic field as the distortion of the relativistic length that arises in traveling around a closed curve. Weyl's interpretation was disputed by Einstein, and was never fully accepted. After the emergence of quantum mechanics with its emphasis on wave functions, however, it became clear that phase rather than scale was the correct concept for Maxwell's equations, or to use modern terminology, that the gauge group was the circle rather than the positive real numbers. Unfortunately, though it was easy to fit alterations of scale into Einstein's theory by replacing the metric with a more flexible structure, there was no room to incorporate phase into general relativity theory. Gauge theory should be added rather as an overlying structure onto the space-time manifold, and thereby the unification Weyl was seeking vanished.

The circle group is commutative, or *abelian*, as one says in tribute to the great Norwegian mathematician Niels Henrik Abel (1802–1829), who was one of the pioneers of group theory. This simply means that a multiplication of elements in the group is independent of the order in which the multiplication is carried out ($a \cdot b = b \cdot a$). In the gauge theory, which Weyl established, it is the circle group that is the gauge group, and we speak therefore of an *abelian gauge theory*. This theory, as we have just said, is inadequate. *Non-abelian gauge theories* were introduced in 1954 by Yang and Mills, and these theories have been intensively studied by physicists and mathematicians ever since. Today, it is among the liveliest areas of research in mathematical physics, and it is not only the mathematics that has had crucial significance for the development of the physical theory; the influence has also gone in the other direction, as we see in the trailblazing work of Simon Donaldson, who used gauge theories in the study of 4-manifolds, work that earned him—as noted in Chapter 2—the Fields Medal at the International Congress of Mathematicians in Berkeley in 1986.

The connection between gauge theories and the theory of fiber bundles was ignored or regarded as irrelevant until the mid-1970's. Since then, we have experienced one of the fruitful symbioses between mathematics and physics with which the history of science is so rich. In what follows, we shall briefly describe the connection between physics and geometry in a gauge-field theory.

Let us imagine a particle with structure, that is, a particle placed in a point x of the space-time manifold M and having an internal structure, or a system of states described by the elements g in a group G. The group could be the cyclic group of order 2, if we wished to describe spin, or the circle group, if we should describe phase. The copy of the group G that describes the internal structure of the particle in the point x of M is denoted by $G(x)$. We then consider the total space E of all possible states for such a particle. The internal spaces $G(x)$ and

$G(y)$ lying over two different points x and y in M, are regarded here as being different. In a natural way, E is a fiber bundle over M with $G(x)$ as fiber over x.

When there is no external field, we can regard all the spaces $G(x)$ as being identified with each other, such that in this situation the fiber bundle is the trivial fiber bundle with base space M and fiber G.

Now let us imagine that the system is placed in an external field that has the effect of distorting the trivial arrangement of the fibers such that an immediate identification of the different spaces $G(x)$ in different points is no longer possible. We assume, however, that the spaces $G(x)$ and $G(y)$ can still be identified if we choose an arbitrary but fixed path in M from x to y. (The identification depends on the chosen path.) In more physical terms, we imagine that the particle moves from x to y and carries its internal state with it. (In Minkowski space, such a motion will take place along the world line for the particle.) The identification of the fibers occurs by a parallel transport. If we now imagine two different paths connecting x and y, then there is no reason to believe that the two different parallel transports derived from these paths agree with each other; we assume therefore that they deviate by a multiplication of a group element, which can be interpreted physically as a generalized phase shift. Mathematically, this element is regarded as the total curvature, or distortion, of the fiber bundle over a suitable surface segment in the base space bounded by the two paths.

If we now consider this picture on a small scale, we can define an infinitesimal (the mathematical word for an arbitrarily small) parallel transport in a point x in a given direction. This describes the infinitesimal shift A of the fiber $G(x)$ into nearby fibers, and this is called a *connection*. The infinitesimal curvature F is dependent on two directions in x and corresponds physically to an infinitesimal phase shift. The infinitesimal picture, that is, the connection, can be summed incrementally (by integration) to give the global picture of the parallel transport along curves such that the two viewpoints are mathematically equivalent.

If we now compare this picture with the situation where there is no external field and where all the fibers $G(x)$ are coherently (i.e., continuously) identified, then we can regard parallel transport as a shift of phase in a fixed copy of the group G, and the connection as what the physicists call the *gauge potential*. In the same way, the curvature F corresponds to the *gauge field*. The curvature F can thereby be construed as the distortion produced by the external field, or it can be identified with the field, when we think of a field as a force measured by its local effect. This identification of fields with geometric distortions is of course also entirely central to Einstein's theory of gravitation. The difference is that the distortion here does not take place in the geometry of the space-time, but in the geometry of an overlying fictitious state space of the space-time. This difference renders the relevant geometry less transparent, and the geometry of fiber bundles also appeared historically later than the geometry of space, both in physics and in mathematics. It is interesting that both mathematicians and physicists, each for their own motives, were led to study these objects that actually crop up in many other completely different contexts.

Despite its historically late appearance, the geometry of fiber bundles of the type we have described here is, from a technical point of view, far simpler than the geometry of the spaces that were used by Riemann and Einstein. This is because the groups that appear as state spaces in gauge theories are

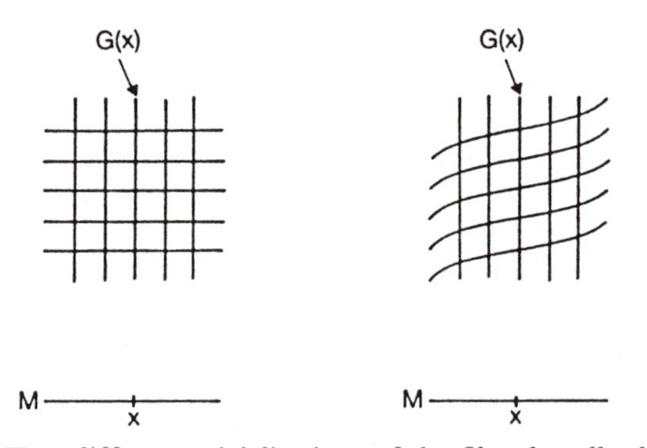

Figure 5.22. Two different trivializations of the fiber bundle that describes a system without an external field.

simpler than the groups in Riemannian geometry, where the group of all local coordinate shifts has to be taken into consideration.

To describe the geometric parallel transport (the connection) more closely, we compare it to the case without an external field. In this case, we employed a coherent identification of all the fibers $G(x)$, but it is important to emphasize that the coherence is due to the absence of an external field, and that the actual choice of a coherent identification (trivialization of the fiber bundle) is left to us. A particular choice is called singling out a *gauge*, and a change from one choice to another is called a *gauge transformation*. As shown in Fig. 5.22, we imagine two different trivializations described by two different sets of "horizontal sections" in the fiber bundle, and the change from the one set to the other is described by a function $g(x)$, with values in the group G. The picture notwithstanding, there is not one particular choice that can be regarded as preferable, and only when a gauge has been chosen (singled out) can the connection and the curvature be indicated. In the case of a system with an external field, the set of local gauge transformations (defined on charts in the space-time manifold M) plays a role analogous to shifts among different charts on a smooth manifold; they piece the global connection together. Since it is a simpler group that appears, the geometry of fiber bundles is an easier theory—it is, in a precise sense, less nonlinear.

Gauge transformations are familiar from classical electrodynamics. Here, they express the fact, for example, that in Maxwell's laws we are free to add the gradient* of a differentiable function to the magnetic potential, since the curl operator **curl** vanishes on a gradient field. Gauge transformations also appear in connection with phase shift for a wave function that describes an elementary particle. Such phase shifts are functions $\theta(x)$ with values in the circle group, and x is a point in the space-time manifold.

It must be emphasized that a connection is a fixed geometrical object, and that it is more primitive than curvature. A consequence of this is that the gauge potential is to be regarded as more primitive than the gauge field. For many years, it was thought that the electromagnetic potential was merely an artificial mathematical quantity that was useful, but could not be assigned any physical meaning. In 1959, however, Yakir Aharonov and David Bohm proposed an experiment, first carried out by Robert G. Chambers in 1960, revealing that even in the absence of a field, the electromagnetic potential plays a role. In this experiment, a uniform beam of electrons

* *The gradient field* of a differentiable function $\theta(x)$ is a vector field that in every point x indicates the direction in which increase of the function is greatest. The magnetic field vector appears by applying the curl operator to the magnetic (vector) potential.

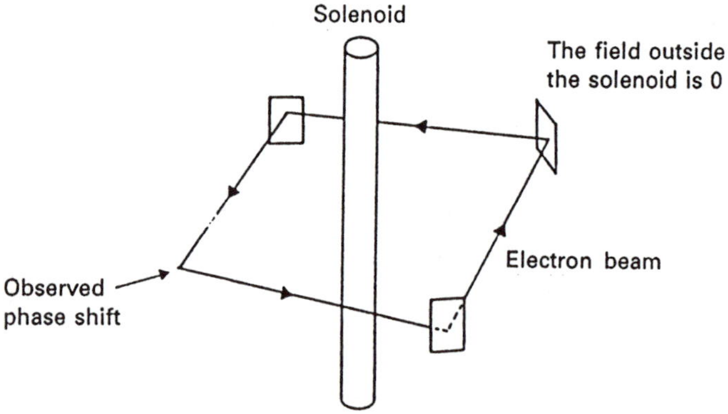

Figure 5.23.

is deflected in a closed path around a coil. (See Fig. 5.23.) The coil is regarded as an infinitely long, perfectly isolated tube. Although the field outside the tube is zero, a phase shift can be observed occasioned by the beam's action on itself (interference), and this phase shift varies with the strength of the current in the coil. Many experimental physicists criticized Chambers's experimental arrangement, but finally in 1986, a corroborating experiment was carried out by a research group led by Akira Tonomura. Important technological applications are expected from the Aharonov–Bohm effect in microelectronics. See, *inter alia*, the article "Quantum Interference and the Aharonov–Bohm Effect," by Y. Imry and R. A. Webb in *Scientific American*, Vol. 260, No. 4. (April, 1989), pp. 36–42.

Mathematically, the Aharonov–Bohm effect can be construed as the fact that even if the field is identically zero, physical effects can nevertheless occur, because the parallel transport need not be trivial if there are "holes" in the relevant area of space. That the curvature is zero gives information only about parallel transport around very small closed paths. In physical terminology, parallel transport is usually described by speaking of non-integrable phase factors. Locally, non-integrability is associated with the field's not being zero, whereas global non-integrability is of a topological character (e.g., going around a wire), and can appear even for zero fields (outside the wire).

Classically, potentials were introduced as a mathematical aid in simplifying the field equations, and the freedom of choice (or gauge freedom) of the gauge potential was taken as an expression that the potential did not have any genuine physical significance. The geometric viewpoint shows, however, that this is too narrow a conception. The connection is a geometrical object, and therefore the potential also ought to be conceived as a physical quantity. The nonphysical has to do with the choice of gauge, where one chooses to describe the potential, corresponding to the fact that the geometrical fiber bundle that carries the connection does not *a priori* have any natural horizontal cross sections.

We may also speculate as to whether the principal fiber bundles, which model field theories, are topologically trivial over topologically non-trivial space-time manifolds. In 1930, Dirac introduced the concept of a *magnetic monopole* as an electromagnetic field with an isolated singularity in space. He calculated that the integral of the field over a sphere around the singularity could assume integral values different from zero. Such integrals correspond in mathematics to some characteristic classes of principal bundles introduced by the prominent differential geometer Chern in 1946. These magnetic monopoles have not to date been detected experimentally. There may be a Nobel Prize in physics waiting to be won here.

As stated in the introduction to this chapter, gauge theories were introduced with the intention of unifying the four fundamental forces in nature: gravity, the electromagnetic force, and the strong and the weak interactions. For every group G, we can construct a gauge theory, and the problem then is to make a choice of group that corresponds as closely as possible to the physical observations. Gravity is the problem; for the other three forces, we now seem to have a good gauge theory. The formulation of quantum-electrodynamics (which couples electromagnetism with quantum mechanics) as a gauge theory with the circle group as group was carried out by Hermann Weyl in 1929. In a great achievement at the beginning of the 1970s, honored by a Nobel Prize, Salam, Glashow, and Weinberg unified electrodynamics and the weak interactions in a gauge theory with the group U(2) as gauge-group. To further incorporate the strong interactions, Georgi and Glashow at the beginning of the 1980's developed a gauge theory with the still larger group $SU(3) \times U(2)$ as gauge-group. The incorporation of gravity has remained more elusive, so the physicists search for other theories to reach that desired goal.

String theory is now regarded by physicists as capable—with a little luck—of simultaneously quantizing gravity and the other natural forces. In a string theory, the fundamental objects are no longer point-shaped particles moving along curves in a space-time manifold, as they do in field theory. Now they are open or closed strings that sweep out a two-dimensional surface, called the *world surface*, as they travel through space time. In contrast to an ordinary field-theory, where different forms of interaction can be postulated, there is no possibility of choice for interaction in a string theory, because the action in this theory is purely geometric and a topological characteristic of the world surface. Among the inescapable demands on a string theory is that it dictates a space-time manifold of a particular (and high) dimension through which the strings travel; 10 dimensions for the so-called superstrings and 26 dimensions for bosonic strings. The research in string theory is on a very high level of mathematical abstraction, and we must refrain here from proceeding into these theories.

Among the mathematicians who have contributed to the study of gauge theories, we must single out Sir Michael Atiyah, who established a school for this research at Oxford University. From this school Simon Donaldson, among others, has emerged. Sir Michael has over the years made several important contributions to mathematics, and in 1966 he was awarded the Fields Medal for his work in topology and analysis. It will take us too far afield to mention other names here, for the field is currently extremely active.

EPILOGUE

We have reached the end of the road; for now, our journey through the world of geometry is over.

We hope it has succeeded in demonstrating the vital role geometry plays in our conception of the world around us, and has offered insights into phenomena that do not yield up their secrets at first glance. Geometry describes the basic forms and figures of which the environment is composed, and image formation is a fundamental element in the comprehension of this.

One cannot but be impressed by our forefathers. It was a stroke of genius for Eratosthenes to measure the circumference of the Earth—and daring for his time. It was an act of courage as well as genius for Galileo to direct science onto its present course. Newton and Maxwell revolutionized physics, and their discoveries form the basis of modern technology. Einstein's relativity theory, which is intimately tied to geometry, and the theory of quantum mechanics, describing the world on the atomic level, have been of far-reaching significance for technological conquests in our century.

Discoveries have also been made in connection with the development of the modern field theories—inseparable from modern geometry—that can have crucial significance for mankind.

It is striking that most discoveries have led to technological advances only at some later—sometimes much later—time. Disciplined impatience ought therefore to be the hallmark of science.

Geometry strengthens imagination and creativity, and together with the other mathematical disciplines, geometry is a part of the foundation of modern technology. It is my hope that the reader has realized that mathematics, too, is part of our cultural heritage, and has perceived our ties to the mathematicians of the past, deriving pleasure from arguments that—though they come to us from long ago—have retained their sharpness and beauty.

In parting, let me express the hope that the reader has experienced the same pleasure in reading the book that I have had in writing it. Enjoy your continued journey through the world, which—remember—has a geometric dimension.

Bibliography

In the bibliography below we list the most essential sources for the material in each chapter. Mentioned are also a few other books from which the interested reader can learn more about some of the subjects. A few places in the main text special references are given which are not repeated in the list below.

Chapter 1. *Substantial inspiration has been found in the books*:

H. Weyl: *Symmetry*. Princeton University Press, 1952.

E. H. Lockwood: *A Book of Curves*. Cambridge University Press, 1961.

L. A. Lyusternik: *Shortest Paths – Variational Problems*. Pergamon Press, 1964.

R. Osserman: *A Survey of Minimal Surfaces*. 2nd Edition. Dover Publications, New York, 1986.

Among the books on mathematical forms in nature we recommend:

D'Arcy W. Thompson: *On Growth and Form*. Cambridge University Press. 2nd Edition 1942. (The classic over all classics.)

P. S. Stevens: *Patterns in Nature*. An Atlantic Monthly Press Book. Little, Brown and Company. Boston/Toronto, 1974. (Nontechnical, but good.)

S. Hildebrandt and A. Tromba: *Mathematics and Optimal Form*. Scientific

American Library, W. H. Freeman and Company, San Francisco, 1985. (A beautiful and exciting book about forms in nature as solutions to optimization problems.)

To the section on geometry of tiled surfaces we recommend:

E. H. Lockwood and R. H. MacMillan: *Geometric Symmetry*. Cambridge University Press, Cambridge, 1978. (Reasonably intelligible.)

B. Grünbaum and G. C. Shephard: *Tilings and Patterns*. W. H. Freeman and Company, San Francisco, 1987. (The book about this subject.)

Finally, two classics about the splendours of mathematics:

R. Courant and H. Robbins: *What is Mathematics*? Oxford University Press, New York, 1941.

D. Hilbert and S. Cohn-Vossen: *Anschauliche Geometrie*. Springer-Verlag, Berlin, 1932.

Chapter 2. *The exposition given here is mainly based on:*

V. L. Hansen: *Klassifikationen af de lukkede flader*. Mathematical Institute, University of Copenhagen, 1977. (In Danish.)

Similar expositions of the classification of the closed surfaces can be found in the following sources:

M. A. Armstrong: *Basic Topology*. McGraw-Hill (UK), London, 1979.

P. Andrews, "The Classification of Surfaces", *The American Mathematical Monthly*, Vol. 95, No. 9, November, 1988, 861–867.

Alternative expositions of the classification of the closed surfaces can be found in the following books:

H. B. Griffiths: *Surfaces*, Cambridge University Press, Cambridge, 1976.

H. Seifert and W. Threlfall: *Lehrbuch der Topologie*. Chelsea, New York, 1947.

Chapter 3. *The most important inspiration has been found in the original texts:*

R. Thom: *Stabilité structurelle et morphogénèse*, Benjamin 1972. English translation by D. H. Fowler, Benjamin/Addison Wesley, 1975.

E. C. Zeeman: *Catastrophe Theory. Selected papers 1972–1977.* Addison-Wesley, 1977.

Among other books on the subject we recommend:

T. Poston and I. Stewart: *Catastrophic Theory and its Applications*. Pitman Publishing Limited, 1978. (Thorough; many references.)

A. Woodstock and M. Davis: *Catastrophe Theory*. Penguin Books, 1978. (Popular exposition.)

V. I. Arnold: *Catastrophe Theory*. Springer-Verlag, 1984. (A very fascinating booklet.)

Chapter 4. *The source for this chapter is the extremely fascinating book:*

M. Kline: *Mathematics and the Search for Knowledge*. Oxford University Press, New York, Oxford, 1985.

Other recommendable books on these subjects include:

I. B. Cohen: *The Birth of a New Physics*. Doubleday, 1960.

H. Butterfield: *The Origins of Modern Science*. Macmillan, New York, 1932.

O. Pedersen and M. Phil: *Historisk indledning til den klassiske fysik I.* Munksgaard, Copenhagen, 1963. (In Danish.)

O. Pedersen: *Matematik og Naturbeskrivelse i oldtiden*. Akademisk Forlag, Copenhagen, 1975. (In Danish.)

J. Teichmann: *Wandel des Weltbildes. Astronomie, Physik und Messtechnik in der Kulturgeschichte*. (Published in a book series from Deutsches Museum in München.) Rowohlt Tashenbuch Verlag GmbH, 1985.

Chapter 5. *For the history of electromagnetism in section 1, the book by Kline is again the main source.*

The exposition of the theory of relativity is mainly based on two articles:

R. Penrose: "The Geometry of the Universe". Pages 83–125 of L. A. Steen (ed.) *Mathematics Today – Twelve Informal Essays*. Springer-Verlag, 1978.

R. Geroch: "General Relativity". *Proc. Symposia in Pure Mathematics* Vol. 27, 401–414. American Mathematical Society, 1977.

General expositions of the theory of relativity can be found in:

W. Rindler: *Essential Relativity*. D. van Nostrand, New York, 1969; Springer-Verlag, New York, 1977.

W. J. Kaufman: *Cosmic Frontiers of General Relativity*. Little, Brown and Company, Boston, 1977.

E. F. Taylor and J. A. Wheeler: *Spacetime Physics*. W. H. Freeman and Company, San Francisco, 1963, 1966.

Concerning the physics of elementary particles and gauge theories the main sources are the following articles:

C. H. Taubes: "Physical and Mathematical Applications of Gauge Theories". *Notices of the American Mathematical Society*, Vol. 33, No. 5, October 1986, 707–715.

G. 't Hooft: "Gauge Theories of the Forces between Elementary Particles". *Scientific American*, Vol. 242, No. 6, June 1980, 90–116.

H. J. Bernstein and A. V. Phillips: 'Fiber Bundles and Quantum Theory". *Scientific American*, Vol. 245, No. 1, July 1981, 94–109.

Also the introductory section of the following book is a main source:

M. F. Atiyah: *Geometry of Yang–Mills Fields. Leizioni Fermiane*. Accademia Nazionale Dei Lincei Scuola Normale Superiore, Pisa, 1979.

As a general reference for differential geometric physics can be mentioned:

M. Göckeler and T. Schücker: *Differential Geometry, Gauge Theories and Gravity*. Monographs on Math. Phys. Cambridge University Press, Cambridge, 1987.

Elementary expositions of string theory can be found in:

F. David Peat: *Superstrings and the Search for the Theory of Everything*. Abascus, 1992.

B. Sørensen: *Superstrenge-En teori om alt eller intet*. Munksgaard, Copenhagen, 1988. (In Danish.)

A short nontechnical account of string theories can also be found in the stimulating book:

S. W. Hawking: *A Brief History of Time*. Bantam Books, New York, 1988.

INDEX

Other Titles of Interest

from

A K PETERS, LTD

Michael F. Barnsley and Lyman P. Hurd
Fractal Image Compression
ISBN 1-56881-000-8

Wolfgang Boehm and Hartmut Prautzsch
Geometric Concepts for Geometric Design
ISBN 1-56881-004-0

Philip J. Davis
Spirals: From Theodorus to Chaos
ISBN 1-56881-010-5

Geometry Center, University of Minnesota
Not Knot (VHS video)
ISBN 0-86720-240-8

Amos Harpaz
Relativity Theory: Concepts and Basic Principles
ISBN 1-56881-007-5

Iterated Systems Inc.
Snapshots: True-Color Photo Images Using the Fractal Formatter
ISBN 0-86720-299-8

289 Linden Street
Wellesley, Massachusetts
(617) 235-2210
Fax (617) 235-2404